The First Steps of Life

SCIENCES

Biology, Field Director – Marie-Christine Maurel

Xenobiology, Exobiology, Astrobiology, Origins of Life,
Subject Heads – Marie-Christine Maurel and Marc Ollivier

The First Steps of Life

Coordinated by
Ernesto Di Mauro

WILEY

First published 2024 in Great Britain and the United States by ISTE Ltd and John Wiley & Sons, Inc.

Apart from any fair dealing for the purposes of research or private study, or criticism or review, as permitted under the Copyright, Designs and Patents Act 1988, this publication may only be reproduced, stored or transmitted, in any form or by any means, with the prior permission in writing of the publishers, or in the case of reprographic reproduction in accordance with the terms and licenses issued by the CLA. Enquiries concerning reproduction outside these terms should be sent to the publishers at the undermentioned address:

ISTE Ltd
27-37 St George's Road
London SW19 4EU
UK

www.iste.co.uk

John Wiley & Sons, Inc.
111 River Street
Hoboken, NJ 07030
USA

www.wiley.com

© ISTE Ltd 2024

The rights of Ernesto Di Mauro to be identified as the author of this work have been asserted by him in accordance with the Copyright, Designs and Patents Act 1988.

Any opinions, findings, and conclusions or recommendations expressed in this material are those of the author(s), contributor(s) or editor(s) and do not necessarily reflect the views of ISTE Group.

Library of Congress Control Number: 2023942095

British Library Cataloguing-in-Publication Data
A CIP record for this book is available from the British Library
ISBN 978-1-78945-165-8

ERC code:
LS8 Ecology, Evolution and Environmental Biology
　LS8_5 Evolutionary genetics

Contents

Introduction xi
Ernesto DI MAURO

Chapter 1. The Emergence of Life-Nurturing Conditions in the Universe 1
Juan VLADILO

1.1. Defining properties of life 1
    1.1.1. Implications of the defining properties 2
1.2. Life-supporting conditions and environments 5
    1.2.1. Chemical ingredients 6
    1.2.2. Physical conditions 7
    1.2.3. Habitable worlds 9
1.3. Setting the stage for chemistry and life in the Universe 10
    1.3.1. Births of the laws of chemistry 10
    1.3.2. Production of chemical elements 11
    1.3.3. Assemblage of prebiotic molecules 12
    1.3.4. Origin of water 14
    1.3.5. Appearance of rocky planets 15
1.4. The habitable Universe 16
    1.4.1. Circumstellar habitable zones 17
    1.4.2. Galactic habitable zones 19
1.5. Planetary environments suitable for the origin of life 20
    1.5.1. Abiogenesis on planetary surfaces 20
    1.5.2. Abiogenesis in the oceans 22

1.5.3. Implications for the search for life outside Earth 23
1.6. The quest for inhabited worlds 23
1.7. References 24

Chapter 2. Chirality and the Origins of Life 31
Guillaume LESEIGNEUR and Uwe MEIERHENRICH

2.1. Introduction to chirality. 32
2.2. The asymmetry of life. 35
2.3. The origin of homochirality 37
   2.3.1. Stochastic theories 37
   2.3.2. Deterministic theories 38
2.4. Space missions and the search for life and its origins 41
   2.4.1. Rosetta 43
   2.4.2. ExoMars 45
2.5. References 48

Chapter 3. The Role of Formamide in Prebiotic Chemistry 55
Raffaele SALADINO, Giovanna COSTANZO and Bruno Mattia BIZZARRI

3.1. Introduction. 55
3.2. Effect of minerals and self-organization in the prebiotic
chemistry of formamide 57
   3.2.1. Surface catalysis and geochemical scenarios. 57
   3.2.2. Chemomimesis, circularity and thermodynamic niches 59
   3.2.3. Nucleosides phosphorylation 62
3.3. Continuity and mineral complexity. 63
3.4. Energy-driven selectivity. 67
3.5. References 68

**Chapter 4. A Praise of Imperfection: Emergence and
Evolution of Metabolism** 79
Juli PERETÓ

4.1. From Darwin to Jacob: perfection does not exist. 79
4.2. Protometabolic networks. 82
4.3. Enzyme promiscuity and metabolic innovation. 86
4.4. Promiscuity, moonlighting and the essence of life 91
4.5. Acknowledgments. 93
4.6. References 93

Chapter 5. Viruses, Viroids and the Origins of Life 99
David DEAMER and Marie-Christine MAUREL

5.1. How were viruses discovered? A brief history 100
5.2. Viral diversity 101
5.3. Viral structure and function 103
5.4. Viruses and mammalian genomes 106
5.5. Role of viruses in human evolution, health and disease 107
5.6. Viroids may be a link to ancient evolutionary pathways 108
5.7. Origin and evolution of viroids 109
5.8. Conclusion 111
5.9. References 112

Chapter 6. Is the Heterotrophic Theory of the Origin of Life Still Valid? 117
Antonio LAZCANO

6.1. Introduction. 117
6.2. The roaring 20s 118
6.3. Coacervates as models of precellular structures 121
6.4. Precellular evolution and the emergence of cells. 123
6.5. Final remarks: does Oparin still matter? 128
6.6. Acknowledgments. 130
6.7. References 130

Chapter 7. Making Biochemistry-Free (Generalized) Life in a Test Tube. 135
Juan PÉREZ-MERCADER

7.1. Summary 135
7.2. Introduction and background 136
7.3. Laboratory implementation of an artificial autonomous,
and self-organized functional system 140
7.4. More physics and chemistry working together: phoenix,
self-reproduction via spores, population growth and chemotaxis 144
7.5. Discussion and conclusions 152
7.6. Acknowledgments. 153
7.7. Appendices: Some additional emergent features in PISA
"powered" synthetic biochemistry free protocells. 154
    7.7.1. Chemotactic behavior 154
    7.7.2. Adaptive behavior and click-PISA. 155

7.7.3. Competitive exclusion principle and iniferter PISA. 156
7.7.4. PISA and its control by chemical automata. 156
7.7.5. Integrating PISA and information control with the
Belousov–Zhabotinsky chemical reaction 157
7.8. References 159

Chapter 8. Hydrothermalism for the Chemical Evolution Toward the Simplest Life-Like System on the Hadean Earth 163
Kunio KAWAMURA

8.1. Introduction. 163
    8.1.1. Realistic life-like systems on the Hadean Earth 163
    8.1.2. Water in universe 165
    8.1.3. Two-gene hypothesis, minerals and high temperature 168
8.2. Hydrothermal environment for the chemical evolution
of biomolecules 170
    8.2.1. As an energy source 170
    8.2.2. Temperature and pressure 171
    8.2.3. Biochemical interactions. 172
    8.2.4. Minerals and the thermodynamically open system 174
8.3. Hydrothermal methodologies regarding the origin-of-life study. 175
    8.3.1. Technical background of research tools for
hydrothermal reactions. 175
    8.3.2. Recent development using flow system 176
8.4. RNA world versus hydrothermalism 178
    8.4.1. Stability and accumulation of RNA 178
    8.4.2. RNA-based life-like system under hydrothermal
environments 182
8.5. Future outlook and conclusions 185
8.6. Acknowledgments. 186
8.7. References 186

Chapter 9. Studies in Mineral-Assisted Protometabolisms 193
Jean-François LAMBERT, Louis TER-OVANESSIAN and
Marie-Christine MAUREL

9.1. Metabolism, protometabolism and minerals 193
9.2. Adsorption on mineral surfaces 196
    9.2.1. Adsorption mechanisms 196
    9.2.2. Adsorption selectivities 197

9.3. Mineral surfaces and reaction thermodynamics 198
   9.3.1. Minerals as reagents 198
   9.3.2. Concentrating reagents from the solution 199
   9.3.3. Altering free enthalpies of reaction 201
   9.3.4. Platforms to capture free energy from macroscopic
          sources (space gradients and time fluctuations) 202
9.4. Minerals and reaction kinetics: heterogeneous catalysis 204
   9.4.1. Lessons from industrial heterogeneous catalysis 204
   9.4.2. What can heterogeneous catalysts do? 205
   9.4.3. Reaction selectivity 206
9.5. A case study: primordial synthesis of pyrimidines 207
9.6. Conclusion 209
9.7. References 210

Chapter 10. A Rationale for the Evolution of the Genetic Code in Relation to the Stability of RNA and Protein Structures .. 217
Andrew TRAVERS

10.1. Introduction 217
10.2. Codon–anticodon recognition 218
10.3. Concluding remarks 226
10.4. Acknowledgments 226
10.5. References 226

List of Authors 231

Index 233

Introduction

Ernesto DI MAURO
Institute of Molecular Biology and Pathology, CNR, Rome, Italy

Origin of life studies are an active field whose borders are poorly defined. It allows us to approach the problem with intellectual freedom, out of the limitations imposed by sclerotized disciplines. This book proposes fly-over of this large territory and an in-depth eagle-eyed look of the opinions and the results obtained by some among the active contributors of this fascinating and deeply thought-provoking matter. The topics are presented as usual according to a bottom-up order: the habitability of the Universe, the rationale behind meaningful prebiotic chemistry, the possible and the probable prebiotic chemical frames, the problem of chirality, the role of minerals in biogenesis, the biogenic fertile environments, the in-and-out problem as solved by vesicles physics, the way of the codes, LUCA and protometabolisms, the meaning of complex biological biomorphs (read: viroids). The evolution of information and complexity is the background scenario, which accompanies all of this reasoning.

A single standard-sized book provides a limited space, and in this context 10 chapters are a meagre body. Potential, appropriate and relevant contributors could have been and could be many more. Hence, the effort by each of the authors to widen the field of their descriptions, trying to amend the complexity of this dynamic and multifaceted science.

Connecting cosmology with molecular biology may seem arrogant, but it is not so. The length of the road is the precise indication that we are confronted with an enormous problem that can only be addressed with humility. The daring endeavor is to try to unify under the same perspective topics and discoveries made in very different fields, often told in languages which sometimes are difficult to reconcile. The reader will, in the end, realize that a solid synthesis is for the moment lacking but that, at the same time, the thread connecting the Big Bang to our existence is beginning to become traceable.

Another important message conveyed by the pages of this book is that the approach to the origin of life should be as devoid as possible of anthropocentry. The Universe does not really care for the fact that the leaves will fall from the tree of my garden at due time and "*I*" will die. But, at the same time, the Universe takes into consideration that life exists, that life is one of its epiphanies. Here, on this planet and as far as we know, life is a continuous uninterrupted and internally interconnected process. Life is a category of phenomena that can be interpreted only if we look at it with the necessary aloofness.

In order to emerge, life requires stability, simplicity and reactivity. It also entails complexity. The recognition of the necessity of these properties is necessarily accompanied by a series of question marks.

Stability

"Something came from nothing because it was more stable than nothing" (Stenger 2006). The uncontroversial truth contained in this aphorism by Victor Stenger applies to the evolution of the organization of matter in the whole Universe which, for what concerns us here, comes out from the Big Bang. In particular, and even more so, this concept applies also to life. A telling example is provided by polymers which, most of the time, are a stabilization form of the monomers that compose it.

As a side product of this forced and directed tendence to stability, internally repetitive structures are produced, which are thus able to elaborate information. They do so by introducing and selecting small variations. This process, in biological terms, is dubbed evolution. Is this an intrinsic property of polymers? It is so only for some of them, and if so which ones? Only for those which undergo repeated cycles of synthesis/degradation? Variations usually and typically occur within the same class of molecules: one amino acid to another amino acid in a peptide chain, and one nucleic base to another nucleic base in RNA or DNA. Evolution is the evolution of information.

A short-term sort of variation is that occurring in the so-called metabolic cycles. Remaining in textbook examples and in a central gear of the machinery (the Krebs cycle), a compound, say citrate, "changes" under the effect of the environment and becomes cis-aconitate, which becomes D-isocitrate, which becomes alfa-ketoglutarate, then succinyl-CoA, and succinatate, fumarate, L-malate, oxaloacetate, which again becomes citrate. But in the meantime, mediated by the interaction with the environment and with ancillary molecules, something "vital" has occurred and, depending on the direction of rotation of the cycle, the transfer of energy or carbon has orderly entered the system. The prebiotic valence, the determination of the whereabouts of these cycles and the possibility of their reconstruction are becoming experimented reality (Muchowska et al. 2019; Isnard and Moran 2020; Preiner et al. 2020; Yadav et al. 2022).

A long-standing debate has dominated the scene of origin of life studies: "genetics-first" or "metabolism-first"? Occasionally, also "membranes-first" accompanied the bias. The answers to these initially reasonable questions found their well-grounded advocates, but it gradually became clear that no real distinction was possible, and that without a system to harness and control energy (a system that is one of the incarnations of the concept of "phenotype"), no replication and transmission of the genotype might have been possible. With no genotype, the phenotype would have succumbed to the laws of disorder and the domination by local conditions. Hence, the necessity of the simultaneous, concurrent and cooperative evolution of both phenotype and genotype, of both nucleic acids and proteins, and of carboxylic acids, all contained in membrane-defined spaces where concentrations and selections could take place. The prebiotic chemistry involved was thus necessarily large-spectrum, and no solitary, selective and fastidious synthesis would have been likely to win the race.

The history of the evolution of the interaction of the RNA world with the protein world goes through appealing and elegantly solved chapters. The structure of the huge ribosome machinery, as determined by the 2009 Nobel Laureates Venkatraman Ramakrishnan, Thomas A. Steitz and Ada E. Yonath, still conserves in its central RNA core the signs of its initial function of peptide-bond maker. The story of the evolution of ribosomes (Yonath 2010; Belousoff et al. 2010; Petrov et al. 2014; Bose et al. 2022) shows how proteins structurally and functionally gradually replaced RNAs. The extant universal presence of protein enzymes has the same evolutionary history consisting of substitution of the functions initially performed by RNA catalysts. Omnipresent ribo cofactors as NAD are there to remind us. A big part of this story is still unknown. The possibilities of proteins of acquiring independence from nucleic codes are well summarized in Foden et al. (2020) (see also Muchowska and Moran 2020).

I have written above the word "information". Nothing can be more dangerous. The word information is multisemic, like "richness", "democracy" or "beauty". Its meaning depends on the reader and on the context. It does not help much by saying that "information is energy", nor by saying that information theory is a highly developed science, starting from George Boole going all the way to AI. And it does not help quoting in passing that Claude Shannon obtained his PhD in Cold Spring Harbor Laboratory, the temple of American molecular biology, with a thesis on "An algebra for theoretical genetics".

In this Universe, everything is information. If our goal is to define life and to find the borders between the living and the nonliving, we need to mitigate this conundrum by focusing on a selected part of the problem. The information of the extant organisms on this planet encompasses, for instance, epigenetics and topology, which are emergent properties embedded in the linear information of our founding polymers. Epigenetics greedily multiplies genetic information and inscribes on DNA the history it goes through; supercoiling (a typically topological property) is a way to multiply and modulate information to regulate and direct genetic expression and replication; some molecules can handle it, some cannot. Are epigenetics and topology only part of thermodynamics, or are they elaborate and highly evolved properties of living entities? The answer by the layman scientist is as follows: both are emergent properties of living entities. Hence, life entails superposition of information levels, which is one of the keys of the evolutionary process.

The question, "how much in life can be considered an emergent property?" is thus justified. If we are trying to retrace the first steps of life, we need to have this question clear in mind.

Keeping the conservative approach of considering life only the ensemble of the structures and interactions of those phenomena that we know based on our terrestrial experience, the classes of molecules that are interesting for us here are the ones that we have under the eyes: DNA, RNA, proteins, carboxylic acids and aliphatic chains. These molecules are each endowed with their own characteristic properties and, at the same time, are functionally interconnnected. Their mutual connections give rise to cycles, conceived in that form of living topology that owns much to Eörs Szathmary (2000). In this perspective, life is the interplay of multiple cycles, whose wheels turn according to the information sedimented by their history and by the environments they have gone through. The rhizome of life is a web of cycles, as intuited by Deleuze and Guattari (1980). Life is a unitary phenomenon, a property which usually escapes our attention. But life is also multiform, and the interest for its particulars often makes us forget the overview.

According to Deleuze and Guattari, a rhizome follows the principles:

– 1 and 2: *Principles of connection and heterogeneity: every point of a rhizome can be connected to every other and must be.*

– 3: *Principle of multiplicity: only when the multiple is effectively treated as a noun, "multiplicity", then it ceases to have any relationship with the one.*

– 4: *Principle of no-significant breaking: a rhizome can be interrupted, but it will start again on one of its old lines or on new lines.*

– 5 and 6: *Principles of cartography and decalcomania: a rhizome cannot be traced back to any structural or generative model; it is a map and not a track.*

The evocative power of these principles, expressed in a somewhat heterodox way, is interesting. According to these metaphors, the world of the living is a totally connected network in which every single organism is the incarnation of a specific genotype that begins its life at a given moment, determined by the replication/recombination of its parental genetic materials, and ends at the moment of the dissolution of its own specific genome. Each genotype is online with the genotypes from which it derives and with those that could derive from it, it is informationally connected with them. The living network is a unity embedded in space-time and in the genetic space extending, with possible multiple roots, back to the Last Universal Common Ancestor. Before which, by definition, there were only entities immersed in the swamp of combinatorial biochemistry, and only one survivor.

Our body is made up of nucleic bases, amino acids, carboxylic acids and aliphatic chains. The fact that certain classes of molecules (amino alcohols, for instance) do not organize themselves in cycles is particularly interesting and helps to point out the properties which allow other classes of molecules to be dubbed "biogenic". Why? Because life is interaction among polymers, and the polymers interact according to their own structural and functional properties. The physical–chemical environment in which they occur determines who will stay and who will go, and decides whether their properties allow information to form and be transmitted.

Interaction among macromolecules establishes cycles. These cycles have no purpose but to maintain themselves: a mechanistic property, certainly not a finalistic one. No memory exists of the cycles that do not maintain themselves. It may seem that living entities, "we", are those coded and transmitted cycles, apparent macro-cycles made of life and death. This is true only superficially and partially: death exists for each individual; it is an intrinsic property of each specific and individual cluster of metabolic cycles. But life has not been interrupted since its beginning, by definition. Life and death do not belong to the same category of phenomena.

Stability is the underlying property of all this. The borders of life are the borders of the stability of its constituents and the stability of the information that organizes them.

Simplicity

The materials life is made of are simple and chip in the market of this Universe. Given the time and energy, atoms combine and, as far as we know, their reactions follow the same rules all over the Universe, up until its poorly defined borders. Hydrogen, carbon and nitrogen combine in hydrogen cyanide (HCN). According to well-established observations of the inter- and circum-stellar space (Millar 2015), this compound is the most abundant molecule among those with three atoms that contain a carbon atom; three-atom molecules are already considered molecules with relevant initial level of information. Other different three-component molecules are possible, but they are there in lower amounts, or are not there at all.

HCN is very reactive. The most abundant three-atom molecule not containing carbon is water (H_2O) (note that the fourth most abundant atom oxygen (O) has here entered the combinatorial game) (Millar 2015). Reaction of HCN with H_2O affords NH_2COH formamide. Formamide is thus one of the possible stabilization forms of the energy initially contained in HCN. Formamide is a highly versatile source of potentially biogenic compounds (Saladino et al. 2001, 2012a, 2012b) and provides an easy way to further levels of chemical complexity.

Parallel chemistries to that of formamide are, among many, those centered on HCN, formaldehyde CH_2O, methanol CH_3OH, ammonium formate NH_4^+ $HCOO^-$ and, markedly, on formic acid HCOOH (Mohammadi et al. 2020). Their relevance to plausible prebiotic scenarios depends on the attention dedicated to particular parts of the scene (i.e. the chemistry of sugars, aliphatic compounds and proteins) and on the attention given to particular environments and sets of physical–chemical conditions. The formation of formamide from formic acid and ammonia, for instance, depends on the temperature (Kröcher et al. 2009); these 1C-atom compounds are easily interconverted.

As a matter of fact, formamide seems to be the most versatile and the most adaptive to a large variety of environments, especially the prebiotically plausible ones. In the Urey-Miller experiment, in addition to amino acids and formic acid (Miller 1953), formamide is produced in the boro-silicate glass container (Criado et al. 2021), as expected (Saitta and Saija 2014). The possibility of transformation of each of those compounds into another is facile; it depends, in the endless game of reactions occurring in the sky and on Earth, on where and when we are looking at. The relevant fact remains that our body is made up of H, C, N and O, which are the four most abundant elements.

HCN chemistry is very fertile. From the initial observations of the synthesis of adenine (Oró 1961), to more recent reports (Powner et al. 2009; Sutherland 2016; Sutherland 2017 and references therein), HCN has shown its prebiotic worth, both in the emergence of genetic polymers and proto-metabolism (Yadav et al. 2022).

I personally do not consider HCN chemistry as being alternative to formamide chemistry, nor do I see these two chemistries as mutually exclusive. It depends on the frame of reference in which one considers them. Ultimately, it depends on the prebiotic scenario in which the prebiotic pool formed. Darwin (1871) was, 150 years ago, the first to invoke a warm little pool as the shrine of all biochemistries, but he did not detail its composition, he just had no data for further elaboration of his intuition.

The fact that formamide seems to be the most versatile and the most adapt to a large variety of environments, works as a stabilizing agent of HCN upon its interaction with water, is liquid between 4°C and 210°C, reacts with all sorts of possible catalysts yielding large and complex mixtures of prebiotic compounds under all sorts of energy sources (Saladino et al. 2001, 2012a, 2012b) points to the possible universality of its function and to its presence in the imaginary Darwin pool. The versatility of formamide in accepting as catalyst every mineral tested, from the most common terrestrial oxides to a large variety of meteorites, extends the interest of the function of minerals in prebiotic chemistry. The pivotal role of boric acid (H_3BO_3) in the prebiotic chemistry of the pentose moiety was shown (Prieur 2001).

About a definition of life

This book does not have the presumption to describe life in all its relevant aspects. Nevertheless, we should at least have a precise idea of what we are talking about. To define life is very difficult, it can be very accurately described, but its formal definition is elusive. The starting point can only be the definition by Schrödinger (1992) who considered life as that which "avoids the decay into equilibrium".

Edward Trifonov deduced the consensual definition of what life is: "life is reproduction with variations". This definition does not establish what life is, but provides the consensual definition accepted by contemporary science. The definition was obtained by carrying out a comparative analysis of the 123 existing definitions of the word "life" (Trifonov 2011), starting from the consideration that certain words are more represented than others, thus possibly allowing us to reach a consensus. The structuralistic method employed consists of an initial grammatical analysis, followed by a grouping and by a frequency analysis.

Another good, although less stringently obtained, definition is: life is a self-sustained chemical system capable of undergoing Darwinian evolution (Joyce 1994). This definition has not met, since its 1994 formulation, with major objections, even if a formal analysis reveals some uncertainties. Life does not support itself, as it absorbs and processes energy from the outside; it is not a system, but rather a process; and defining a process not for what it is but for the fact that it can change (i.e. evolve: "Darwinian evolution") is, in terms of logic, a weakness. Furthermore, we can imagine an environment in which there are no variations well (i.e. in which changes from an optimized status quo can only be harmful); or in which the variations are cyclical and very high frequency, or that they occur in a time scale that does not correspond in a commensurable way to the time scale of the living entities it hosts and supports. In such environments, evolution could not be a categorical property.

A definition worth quoting is that by Emile Cioran: "Life is the kitscht of matter, ..., it is rupture, heresy, derogation from the rules of matter" (Cioran 1960). This is not scientific reasoning, but it highlights by reductio ad absurdum an important aspect of the problem: life is not an emergent property of matter, life is well within its rules. How could we not appreciate these words? A perfect crystal, a diamond for example, does not depart from its elegant rules of symmetry, but it does not live. Life is complexity, an application of intricate and codified rules.

My preference goes to Trifonov's because it clearly defines the *consensus* of what science at large thinks that life is. The other formulations aspire to be absolute definitions, but each one is only one among the other 122 alternatives.

Reactivity

For life to emerge, a balance between reactivity and stability is necessary. A clear example of the need for this balance is inside us: RNA. Our >4 billion nucleotides genome is made up of DNA, which is a very complex set of molecules. It has not arisen to its complexity magically and abruptly; it has been derived from the simpler, more reactive and more unstable cognate molecule RNA. Everywhere in the biological world both DNA and RNA are at present made by sets of enzymes codified in RNA or in DNA themselves, which are molecules more complex than their products. But this cannot have always been the case. At one point, the abiotic (life did not exist yet), nonenzymatic (enzymes did not yet exist), spontaneous generation of RNA had to occur. This can be retraced, even though not in detail, at least in principle.

Nonenzymatic polymerization of ribonucleotide monomers is observed when the precursors are activated as phosphorimidazolides or nucleotide triphosphates (as reviewed in Dorr et al. 2012), or are in the 2', 3' or 3', 5' cyclic form (Costanzo et al. 2009; Costanzo et al. 2021; Wunnava et al. 2021). Phosphorimidazolides and nucleotide triphosphates are not very likely to be abundant prebiotic compounds, due to the complexity of their synthesis and to their high reactivity and subsequent instability. Nevertheless, they have been very useful in the characterization of abiotic polymerization reactions. Nonenzymatic syntheses of RNA have a high-standing record of results, leading to the understanding of the properties of this molecule and its evolutionary capacities. Appropriate points of entry in the vast literature of this topic are mentioned in some previous studies (Mariani et al. 2018; Walton et al. 2019; Kristoffersen et al. 2022).

Focusing our attention to cyclic compounds, 3', 5' cyclic nucleotides possess sufficient stability to withstand the extreme conditions most likely present on the early Earth. 3', 5' cGMP, in particular, has unique aggregation properties which lead to the formation of oligonucleotide sequences (Costanzo et al. 2009, 2021; Šponer et al. 2021 and references therein), also in acidic (Wunnava et al. 2021) conditions. Nucleotides containing a phosphodiester linkage confined in a strained five-membered ring are stable in certain conditions and are unstable in others; they only require moderate activation energy for undergoing to ring-opening reaction and polymerization. The basic requirement for polymerization is the previous formation of an ordered structure, and these cyclic forms seem to be made on purpose for this. Cyclic CMP affords very short oligonucleotides (Costanzo et al. 2017). The cyclic forms of the other prebiotically potentially relevant nucleotides (especially so for 3', 5' cAMP) do so at a lower extent.

The resulting macromolecular RNA is the product of the perfect compromise between stability and reactivity of its precursors. These precursors have the right structure to order themselves (Chwang and Sundaralingam 1974), and the resulting polymer has an effective capacity to establish a balance order/disorder depending on the environment.

The RNA polymer is a trade-off between stability and reactivity. Its O in 2' is there to allow recombination and sequence evolution, but it is also there to determine a half-life and the hydrolytic susceptibility that puts the molecule at the mercy of the environment. Hence, evolution of RNA sequences by internal and external recombination, hence the evolutionary loss of the 2' O resulting in the appearance of stabilized genomes made up of (2'-deoxy) DNA.

The stability principle applies well to the polymers we know. Why DNA? Because it is stable. A recent example of stability in which genetic information is inscribed: in the ancient palace built by King Herod the Great in the first century BC date seeds were found. Sarah Sallon at the Louis L. Borick Natural Medicine Research Center in Jerusalem and her colleagues succeeded in growing seven date palm trees (*Phoenix dactylifera*) from them (Sallon et al. 2020). This would not have been possible if the genetic material was RNA.

The reactive properties of RNA are at the basis of its power to give rise to complex sequence (Guerrier-Takada et al. 1983; Zaug et al. 1983; Cech 1987; Zaug and Cech 1996), as discovered in the ground-breaking observations by Altman and Cech. Interestingly, even simple-sequence spontaneously-generated short ribonucleotides have the property of recombination (Pino et al. 2013; Stadlbauer et al. 2015; Costanzo et al. 2021). We see this intrinsic ability in action in today's ribozymes (Kaddour et al. 2021 and references therein), indicating potential connections with basic mechanisms at the origin of biogenic information. The reactive properties of RNA allow its nonenzymatic ligation (Usher and McHale 1976; Pino et al. 2008, 2011) and the formation of not-templated complex structures (Wu et al. 2022).

Careful examination of the living systems teaches us that all is relative, even the concepts and the systems that at first sight seem to be a one-way road, as the genetic code was long considered. Thus, the questions must always be asked: "why and how?". Why and how do we have these codes for the gears that keep the machine running? The answer is that, again, it is a matter of balance between combinatorial chance and thermodynamic necessity.

The genetic code: are alternative codes possible? Yes. A Letter to *Nature* by Hall (1979) was the first comment to discovered differences to the standard code. Now this seems to be an ascertained and wide-spread reality (Shulgina and Eddy 2022). The code is not universal and unique, alternatives are possible. In the game between chance and necessity, happenstances and their history have an important role. Now we know that the winning code could have been different. And we know that we are here the way we are for *both* chance and necessity.

It is proposed, and largely agreed upon, that the amino acids composing extant proteins were not 20 since the very beginning (Jukes 1963; Crick 1966, 1968; Trifonov 2000, 2001; Travers 2006). The initial protein world was simpler and initially made of a first generation of amino acids, namely, glycine, alanine, proline and arginine (Travers 2022). The codons of these four amino acids are as follows:

– glycine, GGU GGC GGA GGG;

– alanine, GCU GCC GCA GCG;

– proline, CCU CCC CCA CCG;

– arginine: AGA AGG.

In these codons, the first two relevant letters are Gs or Cs. It is presumed that the third letter, the wobbling one, entered later, at the time of expansion of the code into its present form. The amino acids generated upon irradiation of formamide with a beam mimicking the solar wind are exactly glycine, arginine, proline and guanidine (which is a possible component of arginine) (Saladino et al. 2015); as for the amino acids as a class, little quantity is formed in these experimental conditions. It is also worth recalling what was mentioned above: that G-based nucleotides are the ones which oligomerize spontaneously and whose oligos perform terminal swapping with C-based oligos. Even though these reasonings about the origin of the code and these experimental observations are only mere correlations, the possibility that guanine chemistry and its properties played a role in the very beginning of the code is nevertheless suggestive.

The gear toward complexity

Life may be seen as the development of codes and their interactions. The properties of nucleic acids fill up the domains of chemistry, structure, topology and extend them to high refinement and variety. But nucleic acids cannot go beyond their intrinsic limits. The same is true for proteins, aliphatic chains and carboxylic acids. But if one class of molecules, say the nucleic acids, finds the way of programming and controlling the composition of another class of molecules, say of proteins, and learns how to guide them to specific actions, then the limits are bypassed and the properties of each class do not only add each other but, rather, multiply the possibilities. At that point, coevolution starts. We are today witnessing the extreme sophistication that these guided interactions and coevolution have reached. The classes of molecules other than nucleic acids and proteins cannot be considered only as ancillary actors of this evolutionary game, even though they depend on, use, exploit and allow the codes which control the core of the play.

The generation of the genetic code entails the programmation by nucleic acids of defined protein structures. These are made with a purpose whose outcome is worth coding and conserving. The purpose is to reproduce the code and in so doing to generate over again the cycles which orderly harness energy and regenerate themselves and the system.

What interests us here is the multiplication of possibilities, which is the first consequence of the intimate interaction of the coding and the coded. These possibilities are chemical, structural and topological. If we consider that the properties of nucleic acids are largely different from those of proteins, and that by the coding process nucleic acids master to their own purposes the properties of this other class of molecules, the multiplicatory power of this type of double gear becomes clear.

RNA polymerase has evolved in order to better transcribe what it has to transcribe, in the sites, the moment, the amount, the speed and the frequency it has to have. Other examples of the results of the extension-of-properties-through-the-codes fill the textbooks of molecular biology, acting both around the DNA core of the system and in its periphery.

Nucleic acids-to-proteins coding unifies different chemistries resulting, without any finalism, in life. The epigenetic and the neural codes, based on entirely different mechanisms, embody the possibility of connecting the behavior and the flow of life with the genetic code. The complexity of life relies on the coordination capacity of codes.

A long series of events goes from the origin of the universe and from the formation of atoms until us and until the other forms of life that we have not yet encountered. This is a field of study that has its own validity, not determined by what in biogenic space has come before nor from what will come later. The specific aspects of this series of events can be, and must be, studied independently, as single chapters. Attributing particular values to each of them, or to parts of each, considering them only in the perspective of their possible biogenicity would prevent us from evaluating them in their essence, and it risks being an operation, if not devoid of meaning, at least strongly distorted.

A summary, that every reader is anyhow invited to make, is still premature. In this perspective, it would have been appropriate to follow in the presentation of the arguments a top-down direction, not the bottom-up direction that is usually followed in prebiotic compilations, and that is the one that we respect also here. The reason is that no finalism is rationally conceivable, that the choices that have been made among the many possible ones in the domains of complexity can only by understood a posteriori.

The only suitable logic in order to understand these choices and to grasp what life is, is molecular Darwinism.

Time frames

A recent study (Steele et al. 2022) has detected and characterized the products of organic syntheses associated with serpentinization and carbonation on early Mars. The study was performed on the Martian meteorite Allan Hills 84001 (ALH84001) formed during the Noachian period, with an igneous crystallization age of ~4.09 billion years. Complex refractory organic material was found associated with mineral assemblages formed by mineral carbonation and serpentinization reactions. Serpentinization is the aqueous dissolution of olivines (Sleep et al. 2004). That is, the Noachian period (3.9–4.1 billion years ago) was characterized by water–rock interactions and abiotic organic syntheses on Mars. If the single meteorite analyzed for traces of organic syntheses has yielded positive results, the easy bet is that this is not a happenstance, that serpentinization-related organic syntheses were common.

Was planet Earth undergoing the same process of serpentinization and potentially biogenic organic syntheses in the same period? The fact that serpentinization was widespread during the first billion years of this planet is ascertained (as discussed in Garcia-Ruiz et al. 2000), and in fact it still goes on in places as the Cascade Range and the Rift Valley (Saladino et al. 2016). Prebiotic syntheses performed in serpentinization-related conditions, using formamide as precursor, yielded an extremely rich panel of potentially biogenic compounds, amino acids, nucleic bases, aliphatic chains and carboxylic acids included (Saladino et al. 2016, 2019). The implication is that prebiotic organic syntheses were global scale reactions in early planetary ages, both on Mars and, inferentially, Earth.

These results suggest that the conditions required for the synthesis of the molecular bricks from which life self-assembles, rather than being local and bizarre, seem to be universal and geologically conventional. They also lead to the conclusion that the reactions affording potentially biogenic precursors were limited to a first relatively short period. This period is difficult to define precisely but was probably limited on this planet to the first 200–300 million years. After that, the conditions changed and, where they could, complex coded interactions and pre-biology took over.

The other conclusion is that the time of chemistry and the time of biology are different. The time for prebiotic chemistry on this planet was short but sufficient. Chemical reactions are well defined: they are a box of strongly deterministic events that take place if the right conditions are there, they are fast and do not need much time. The time for biology is totally different matter: once started, it has no limits, its combinations are an open system, evolution can go on until the end of time, at its own local and specific pace. Its complex processes adapt, specialize, modify themselves according to the modifications of the environment, develop functional codes and change them when needed. Being different from chemistry, which is intrinsically limited and repetitive, life endlessly invents variants.

This book aims to explore the overlapped borders between the two categories of phenomena.

References

Belousoff, M.J., Davidovich, C., Zimmerman, E., Caspi, Y., Wekselman, I., Rozenszajn, L., Shapira, T., Sade-Falk, O., Taha, L., Bashan, A. et al. (2010). Ancient machinery embedded in the contemporary ribosome. *Biochemical Society Transactions*, 38, 422–427.

Bose, T., Fridkin, G., Davidovich, C., Krupkin, M., Dinger, N., Falkovich, A.H., Peleg, Y., Agmon, I., Bashan, A., Yonath, A. (2022). Origin of life: Protoribosome forms peptide bonds and links RNA and protein dominated worlds. *Nucleic Acids Research*, 50, 1815–1828.

Cech, T.R. (1987). The chemistry of self-splicing RNA and RNA enzymes. *Science*, 236, 1532–1539.

Chwang, A.K. and Sundaralingam, M. (1974). The crystal and molecular structure of guanosine 3', 5'-cyclic monophosphate (cyclic GMP) sodium tetrahydrate. *Acta Crystallographica*, B(30), 1233–1240.

Cioran, E. (1960). *Histoire et utopie*. Éditions Gallimard, Paris.

Costanzo, G., Pino, S., Ciciriello, F., Di Mauro, E. (2009). Generation of long RNA chains in water. *Journal of Biological Chemistry*, 284, 33206–33216.

Costanzo, G., Giorgi, A., Scipioni, A., Timperio, A.M., Mancone, C., Tripodi, M., Kapralov, M., Krasavin, E., Kruse, H., Šponer, J. et al. (2017). Non-enzymatic oligomerization of 3'-5' cyclic CMP induced by proton- and UV-irradiation hints at a non-fastidious origin of RNA. *ChemBioChem*, 18, 1535–1543.

Costanzo, G., Cirigliano, A., Pino, S., Giorgi, A., Sedo, O., Zdrahal, Z., Stadlbauer, P., Šponer, J., Šponer, J.E., Di Mauro, E. (2021a). Primitive RNA-catalysis with guanine-rich oligonucleotide sequences – The case of a (GGC)$_3$ nonamer. *bioRxive*. doi: 10.1101/2020.05.04.075614.

Costanzo, G., Šponer, J.E., Šponer, J., Cirigliano, A., Benedetti, P., Giliberti, G., Polito, R., Di Mauro, E. (2021b). Sustainability and chaos in the abiotic polymerization of 3',5' cyclic guanosine monophosphate: The role of aggregation. *ChemSystemsChem*, 1(3), e2000057.

Criado-Reyes, J., Bizzarri, B.M., García-Ruiz, J.M., Saladino, R., Di Mauro, E. (2021). The role of borosilicate glass in Miller–Urey experiment. *Scientific Reports*, 11, 21009.

Crick, F.H.C. (1966). Codon-anticodon pairing: The wobble hypothesis. *Journal of Molecular Biology*, 19, 548–555.

Crick, F.H.C. (1968). The origin of the genetic code. *Journal of Molecular Biology*, 38, 367–379.

Darwin, C. (2022). Letters from Darwin, C. to Hooker, J. D. (1 Feb. 1871). Letter, Darwin Correspondence Project/letter/DCP LETT 7471. University of Cambridge, Cambridge.

Deleuze, G. and Guattari, F. (1980). *Capitalisme et schizophrénie. Mille plateaux*. Les Éditions de Minuit, Paris.

Dorr, M., Loffler, P.M.G.L., Monnard, P.A. (2012). Non-enzymatic polymerization of nucleic acids from monomers: Monomer self-condensation and template-directed reactions. *Current Organic Synthesis*, 9, 735–763.

Foden, C.S., Islam, S., Fernández-García, C., Maugeri, L., Sheppard, T.D., Powner, M.W. (2020). Prebiotic synthesis of cysteine peptides that catalyze peptide ligation in neutral water. *Science*, 370(6518), 865–869.

Garcia-Ruiz, J.M. (2000). *Carbonate Sedimentation and Diagenesis in the Evolving Precambrian World*, volume 67. SEPM Society for Sedimentary Geology, Tulsa.

Guerrier-Takada, C., Gardner, K., Marsh, T., Pace, N., Altman, S. (1983). The RNA moiety of ribonuclease P is the catalytic subunit of the enzyme. *Cell*, 35, 849–857.

Hall, B.D. (1979). Mitochondria spring surprises. *Nature*, 282, 129.

Isnard, R. and Moran, J. (2020). À l'origine de la vie : les premières formes de métabolismes sur Terre. *L'Actualité chimique*, 455, 24–30.

Joyce, J. (1994). The foreword. In *Origins of Life: The Central Concepts*, Deamer, D.W. and Fleischaker, G.R. (eds). Jones and Bartlett, Boston.

Jukes, T.H. (1963). Observations on the possible nature of the genetic code. *Biochemical and Biophysical Research Communications*, 10, 155–159.

Kaddour, H., Lucchi, H., Hervé, G., Vergne, J., Maurel, M.-C. (2021). Kinetic study of the avocado sunblotch viroid self-cleavage reaction reveals compensatory effects between high-pressure and high-temperature: Implications for origins of life on earth. *Biology*, 10(8), 720.

Kristoffersen, E.L., Burman, M., Noy, A., Holliger, P. (2022). Rolling circle RNA synthesis catalysed by RNA. *eLife*, 11, e75186.

Kröcher, O., Elsener, M., Jacob, E. (2009). A model gas study of ammonium formate, methanamide and guanidinium formate as alternative ammonia precursor compounds for the selective catalytic reduction of nitrogen oxides in diesel exhaust gas. *Applied Catalysis B: Environmental*, 88, 66–82.

Mariani, A., Russell, D.A., Javelle, T., Sutherland, J.A. (2018). Light-releasable potentially prebiotic nucleotide activating agent. *Journal of American Chemical Society*, 140, 8657–8661.

Millar, T.J. (2015). Astrochemistry. *Plasma Sources Sci. Technol.*, 24, 043001.

Miller, S.L. (1953). A production of amino acids under possible primitive earth conditions. *Science*, 117, 528–529.

Mohammadi, E., Petera, L., Saeidfirozeh, H., Knížek, A., Kubelík, P., Dudžák, R., Krůs, M., Juha, L., Civiš, S., Coulon, R. et al. (2020). Formic acid, a ubiquitous but overlooked component of the early earth atmosphere. *Chemistry, a European Journal*, 26, 12075–12080.

Muchowska, K.B. and Moran, J. (2020). Peptide synthesis at the origin of life. *Science*, 370, 767–768.

Muchowska, K.B., Varma, S.J., Moran, J. (2019). Synthesis and breakdown of universal metabolic precursors promoted by iron. *Nature*, 569, 104–107.

Oró, J. (1961). Mechanism of synthesis of adenine from hydrogen cyanide under possible primitive earth conditions. *Nature*, 191, 1193–1194.

Petrov, A.S., Bernier, C.R., Hsiao, C., Norris, A.S., Kovacs, N.A., Waterbury, C.C., Stepanov, V.G., Harvey, S.C., Fox, G.E., Wartell, R.M. et al. (2014). Evolution of the ribosome at atomic resolution. *Proceedings of the National Academy of Science USA*, 111(28), 10251–10256.

Pino, S., Ciciriello, F., Costanzo, G., Di Mauro, E. (2008). Nonenzymatic RNA ligation in water. *Journal of Biological Chemistry*, 283, 36494–36503.

Pino, S., Costanzo, G., Giorgi, A., Di Mauro, E. (2011). Sequence complementarity-driven nonenzymatic ligation of RNA. *Biochemistry*, 50, 2994–3003.

Pino, S., Costanzo, G., Giorgi, A., Šponer, J., Šponer, J.E., Di Mauro, E. (2013). Ribozyme activity of RNA non-enzymatically polymerized from 3', 5'-cyclic GMP. *Entropy*, 15, 5362–5383.

Powner, M.W., Gerland, B., Sutherland, J.D. (2009). Synthesis of activated pyrimidine ribonucleotides in prebiotically plausible conditions. *Nature*, 459, 239–242.

Preiner, M., Igarashi, K., Muchowska, K.B., Yu, M., Varma, S.J., Kleinermanns, K., Nobu, M.K., Kamagata, Y., Tüysüz, H., Moran, J.M. et al. (2020). A hydrogen dependent geochemical analogue of primordial carbon and energy metabolism. *Nature Ecol. Evol.*, 4, 534–542.

Prieur, B.F. (2001). Étude de l'activité prébiotique potentielle de l'acide borique. *Comptes Rendus de l'Académie des Sciences – Series IIC – Chemistry*, 4, 667–670.

Saitta, A.M. and Saija, F. (2014). Miller experiments in atomistic computer simulations. *Proceedings of the National Academy of Science USA*, 111, 13768–13773.

Saladino, R., Crestini, C., Costanzo, G., Negri, R., Di Mauro, E. (2001). A possible prebiotic synthesis of purine, adenine, cytosine, and 4 (3H)-pyrimidinone from formamide: Implications for the origin of life. *Bioorganic & Medicinal Chemistry*, 9, 1249–1253.

Saladino, R., Crestini, C., Pino, S., Costanzo, G., Di Mauro, E. (2012a). Formamide and the origin of life. *Physics of Life Reviews*, 9, 84–104.

Saladino, R., Botta, G., Pino, S., Costanzo, G., Di Mauro, E. (2012b). Genetics first or metabolism first? The formamide clue. *Chemical Society Reviews*, 41, 5526–5565.

Saladino, R., Carota, E., Botta, G., Kapralov, M., Timoshenko, G.N., Rozanov, A.Y., Krasavin, E., Di Mauro, E. (2015). Meteorite-catalyzed syntheses of nucleosides and of other prebiotic compounds from formamide under proton irradiation. *Proceedings of the National Academy of Science USA*, E2746–E2755.

Saladino, R., Botta, G., Bizzarri, B.M., Di Mauro, E., Garcia Ruiz, J.M. (2016). A global scale scenario for prebiotic chemistry: Silica-based self-assembled mineral structures and formamide. *Biochemistry*, 55, 2806–2811.

Saladino, R., Di Mauro, E., Garcia-Ruiz, J.M. (2019). A universal geochemical scenario for formamide condensation and prebiotic chemistry. *Chemistry – A European Journal*, 25, 3181–3189.

Sallon, S., Cherif, E., Soloway, L., Gross-Balthazar, M., Ivorra, S., Terral, J.-F., Egliand, M., Aberlenc, F. (2020). Origins and insights into the historic Judean date palm based on genetic analysis of germinated ancient seeds and morphometric studies. *Science Advances*, 6. doi: 10.1126/sciadv.aax0384.

Schrödinger E. (1992). *What is Life? With Mind and Matter and Autobiographical Sketches.* Cambridge University Press, Cambridge.

Shulgina, Y. and Eddy, S.R. (2021). A computational screen for alternative genetic codes in over 250,000 genomes. *Computational and Systems Biology, Genetics and Genomes, eLife*, 10. doi: 10.7554/eLife.71402.

Sleep, N.H., Meibom, A., Fridriksson, T.H., Coleman, R.G., Bird, D.K. (2004). H2-rich fluids from serpentinization: Geochemical and biotic implications. *Proceedings of the National Academy of Science USA*, 101, 12818–12822.

Šponer, J.E., Šponer, J., Vyravsky, J., Sedo, O., Zdrahal, Z., Costanzo, G., Di Mauro, E., Wunnava S., Braun, D., Matyasek, R. et al. (2021). Non-enzymatic, template-free polymerization of 3',5' cyclic guanosine monophosphate on mineral surfaces. *ChemSystemsChem*, 3(6), 1–8. doi: 10.1002/syst.202100017.

Stadlbauer, P., Šponer, J., Costanzo, G., Di Mauro, E., Pino, S., Šponer, J.E. (2015). Tetraloop-like geometries could form the basis of the catalytic activity of the most ancient ribo-oligonucleotides. *Chemistry: A European Journal*, 21, 2–11.

Steele, A., Benning, L.G., Wirth, R., Schreiber, A., Araki, T., McCubbin, F.M., Fries, M.D., Nittler, L.R., Wang, J., Hallis, L.J. et al. (2022). Organic synthesis associated with serpentinization and carbonation on early Mars. *Science*, 375, 172–177.

Stenger, V.J. (2006). *The Comprehensible Cosmos: Where Do the Laws of Physics Come From?* Prometheus Amherst, New York.

Sutherland, J.D. (2016). The origin of life – Out of the blue. *Angewandte Chemie, International Edition*, 55, 104–121.

Sutherland, J.D. (2017). Studies on the origin of life – The end of the beginning. *Nature Reviews*, 1, 1–7.

Szathmary, E. (2000). The evolution of replicators. *Philosophical Transitions Royal Society, London B. Biological Science*, 355, 1669–1676.

Travers, A. (2006). The evolution of the genetic code revisited. *Origins of Life and the Evolution of the Biosphere*, 36, 549–555.

Travers, A. (2022). *Why DNA?* Cambridge University Press, Cambridge.

Trifonov, E.N. (2000). Consensus temporal order of amino acids and evolution of the triplet code. *Gene*, 261, 139–151.

Trifonov, E.N. (2011). Vocabulary of definitions of life suggests a definition. *Journal of Biomolecular Structure and Dynamics*, 29, 259–266.

Trifonov, E.N., Kirzhner, A., Kirzhner, V.M., Berezovsky, I. (2001). Distinct stages of protein evolution as suggested by protein sequence analysis. *Journal of Molecular Evolution*, 53, 394–401.

Usher, D.A. and McHale, A.H. (1976). Nonenzymic joining of oligoadenylates on a polyuridylic acid template. *Science*, 192, 53–54.

Walton, T., Zhang, W., Li, L., Tam, C.P., Szostak, J.W. (2019). The mechanism of nonenzymatic template copying with imidazole-activated nucleotides. *Angewandte Chemie International Edition*, 58(32), 10812–10819.

Wu, L.-F., Liu, Z., Roberts, S.J., Su, M., Szostak, J.W., Sutherland, J.D. (2022). Template-free assembly of functional RNAs by loop-closing ligation. *Journal of American Chemical Society*, 144(30), 13920–13927.

Wunnava, S., Dirscherl, C.F., Výravský, J., Kovařík, A., Matyášek, R., Šponer, J., Braun, D., Šponer, J.E. (2021). Acid-catalysed RNA oligomerization from 3',5'-cGMP. *Chemistry – A European Journal*, 27, 17581–17585.

Yadav, M., Pulletikurti, S.R., Yerabolu, J.R., Krishnamurthy, R. (2022). Cyanide as a primordial reductant enables a protometabolic reductive glyoxylate pathway. *Nature Chemistry*, 14, 170–178.

Yonath, A. (2010). Hibernating bears, antibiotics, and the evolving ribosome (Nobel Lecture). *Angewandte Chemie International Edition*, 49, 4341–4354.

Zaug, A.J. and Cech, T.R. (1996). The intervening sequence RNA of Tetrahymena is an enzyme. *Science*, 231, 470–475.

Zaug, A.J., Grabowski, P.J., Cech, T.R. (1983). Autocatalytic cyclization of an excised intervening sequence RNA is a cleavage-ligation reaction. *Nature*, 301, 578–583.

1

The Emergence of Life-Nurturing Conditions in the Universe

Juan VLADILO
INAF – Trieste Astronomical Observatory, Italy

This chapter provides a historical overview of the appearance of life-supporting conditions in our Universe. Throughout this discussion, terrestrial life is regarded as a local example of a phenomenon that may exist elsewhere. The defining properties of life and their implications in terms of molecular properties (section 1.1) are used to infer which chemical ingredients and physical conditions are required for generating a habitable environment (section 1.2). The history of the assemblage of prebiotic ingredients and the formation of rocky planets is then outlined (section 1.3), followed by a description of the current search for habitable worlds and studies of habitable zones (HZs) (section 1.4). Surface and submarine scenarios for the origin of life in planetary bodies are then compared (section 1.5). The final section summarizes the prospects of detecting signatures of life in remote worlds orbiting stars other than the Sun (section 1.6).

1.1. Defining properties of life

In this work, in line with a commonly adopted evolutionary scenario proposed a century ago (Oparin 1924), life is assumed to be the outcome of a gradual increase in complexity and functionality of molecular structures. The progressive transition from the abiotic world to life is assumed to be driven by mechanisms of natural selection operating on the products of a series of spontaneous chemical events.

To introduce the discussion on life-nurturing conditions, it is appropriate to start with a definition of life. Finding a commonly accepted definition is almost impossible (Lazcano 2008). Here, two properties are used to define a living organism as follows:

1) reproduction with variations;

2) synthesis and maintenance of internal constituents.

These properties provide a concise description of *the internal processes, driven by an internal set of instructions, adopted here to define life*. Property (1) is a synthesis of the numerous definitions proposed in the literature (Trifonov 2011): reproduction is a hallmark of life, and variations in the offspring open the door to natural selection and Darwinian evolution. Owing to property (1), life can be seen as a phenomenon able to perpetuate and adapt to long-term environmental changes. Property (2) is another hallmark of life, sometimes called autopoiesis (Varela et al. 1974; Luisi 2016). This property describes, albeit in a schematic way, the instantaneous behavior of an organism during the time span of its existence. From a different perspective, property (1) emphasizes the genetic aspects of life, whereas property (2) emphasizes the metabolic aspects and, implicitly, the need for compartmentalization (see below).

Even though the above definitions have been extracted from the properties of terrestrial life, they do not make any hypothesis on the specific nature of the internal constituents or processes which, in principle, could be different from the terrestrial ones. In this sense, terrestrial life can be regarded as a local example of a universal phenomenon that awaits discovery outside Earth.

It is important to note that life may also exist in an inactive state of suspended reproduction or metabolism. However, the implications discussed below hold for life with active reproduction and metabolism.

1.1.1. *Implications of the defining properties*

In the early steps of life, when life processes are dominated by interactions between molecular structures, properties (1) and (2) bear the following implications:

a) The synthesis and reproduction of internal molecular structures require energy and selected chemical ingredients taken from the environment.

b) The internal entropy has to be kept low because the internal molecular structures need to be arranged in a specific order. In order not to violate the second law of thermodynamics, entropy must be released into the environment in some form (e.g. heat and discarded products).

As sketched in Figure 1.1, the concept of habitability is largely based on (a), whereas the concept of biosignatures (section 1.6) is based on (b). In this scheme, compartmentalization within an open border must exist. The existence of compartments is another hallmark of life (see Luisi 2016) that, for the sake of brevity, is not discussed here.

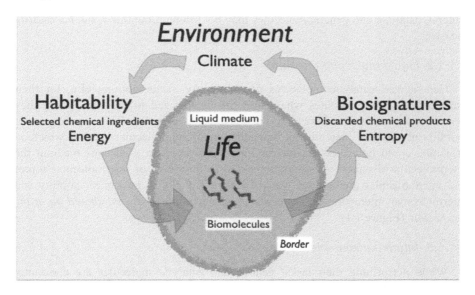

Figure 1.1. *Scheme of the relationship between a living organism and its surroundings*

COMMENT ON FIGURE 1.1.– *The synthesis of internal molecular constituents requires energy and selected chemical ingredients taken from the environment. Since the internal entropy must be kept low, entropy has to be released in some form (e.g. discarded products and heat). Life creates a chemical disequilibrium between the selected and discarded compounds, altering the ambient chemical composition. In the presence of an atmosphere, the gradual changes of atmospheric composition affect the climate and habitability.*

1.1.1.1. Molecular functionalities

The defining properties of life that we have adopted imply the existence of internal molecules that store the instructions for synthesizing and recycling the molecular constituents. These molecules with informational content – hereafter called genetic molecules – should be transmitted to the next generation in the act of reproduction. Moreover, in order to synthesize or recycle molecular structures, and in order to replicate molecules during reproduction, internal molecules with catalytic capabilities – hereafter called catalytic molecules – should also exist. In other words, *the defining properties (1) and (2), when transported at the molecular level, imply the existence of internal molecules with catalytic and genetic functionalities*. In general, catalytic and genetic molecules may be different, but this is not necessarily the case[1].

1.1.1.2. Liquid molecular medium

Genetic and catalytic molecules need mobility, if for nothing else, for their transportation from the sites where they are synthesized to the sites where they perform their tasks or where they are recycled or expelled. The requirement of mobility implies that they must be embedded in a fluid molecular medium which, in principle, could be either gaseous or liquid. However, in a gaseous medium the intermolecular distances are large compared to the range of intermolecular forces and intermolecular interactions would be inhibited. So, to guarantee mobility and intermolecular interactions, *a molecular medium should exist and should be in the liquid state* (Figure 1.1).

1.1.1.3. Intermolecular interactions

While performing their tasks, genetic and catalytic molecules are constantly involved in intermolecular interactions. To preserve their functionalities, *the interactions in which genetic and catalytic molecules are involved should not alter their covalent bond structure*. Therefore, intermolecular interactions should be mediated by chemical forces that (i) do not establish new covalent bonds and (ii) have lower dissociation energies than covalent bonds. The only chemical bonds/forces that satisfy these requirements[2] are van der Waals (vdW) forces and hydrogen bonds (HBs). During molecular interactions, vdW forces are always present, while HBs can be formed only with specific atomic configurations

---

[1] The RNA world can be seen as a special case in which a single molecular structure features both genetic and catalytic capabilities.

[2] Halogen bonds have similar properties but, at variance with hydrogen bonds, they do not provide proton transfer; moreover, halogen elements are rare in the cosmos (see Vladilo and Hassanali 2018).

(see Note 1.1). However, HBs need to be present in life molecular processes because, as is now explained, they have some key properties (Sesiraju and Steiner 1999) that are not present in vdW forces.

Compared to vdW forces, HBs are stronger, more directional and decay more gradually with increasing intermolecular distance (see Tables 2 and 3 in Vladilo and Hassanali 2018). Owing to these properties, HBs have an orienting effect even when two molecules start to approach each other from a relatively large distance. These characteristics, not provided by vdW forces, are crucial for the mutual recognition and correct orientation of two molecules that are starting to interact. The properties of HBs are also essential for the hydrophobic effect, which provides structural support to the genetic and catalytic molecules and is crucial for the spontaneous formation of compartments, such as the border sketched in Figure 1.1. The hydrophobic effect is caused by entropy variations of the molecules of the medium and requires weak, directional interactions. These can be provided by directional HBs, but not by vdW forces, which are nearly isotropic. A further, unique property of hydrogen bonding is its ability to exchange the proton between the donor and acceptor molecules. This exchange enables proton transfer, which is essential for acid–base chemistry reactions: in the absence of proton transfer, it would be impossible to transfer electrons while keeping electrical neutrality in biological systems.

For the above reasons, we expect that *an essential part of the intermolecular interactions of genetic and catalytic molecules need to be mediated by HBs* (i.e. not simply by vdW forces). The same is true for *intramolecular* bonding, not discussed here for the sake of brevity (see Vladilo and Hassanali 2018). These expectations hold for any form of life that satisfies the defining properties (1) and (2) specified above.

NOTE 1.1.– A hydrogen bond (HB) is an attractive interaction between a hydrogen atom from a molecular group X–H, in which the element X is more electronegative than H, and an atom or group of atoms, A, that possesses at least one electron-rich region (Desiraju and Steiner 1999). Due to the difference in electronegativity, the proton in H is partially unshielded and the group X–H (the "donor") is polarized. The electron-rich region in A (the "acceptor") attracts the partially unshielded proton.

1.2. Life-supporting conditions and environments

The definitions and implications discussed in the previous section allow us to infer which chemical ingredients and physical conditions are required to support life. By definition, habitable environments must satisfy these requirements.

1.2.1. *Chemical ingredients*

1.2.1.1. *The molecular medium*

Water has unique properties (see Westall and Brack 2018) that make it essential for terrestrial life. The question is whether or not other forms of life may use an alternative medium. To cast light on this issue, the following requirements should be considered. First, to maximize the efficiency of biomolecular processes, the liquid medium should actively interact with genetic/catalytic molecules, rather than being just a filler. Moreover, the medium should be able to trigger the generation of compartments making use of the hydrophobic effect[3]. Both requirements can be satisfied by a molecular medium able to form a network of HBs. A further requirement is that the molecules of the medium should be much smaller than genetic/catalytic molecules to facilitate their mobility and to fill and support their complex 3D contours. Water molecules are small and can form a complex network of HBs, as each molecule has two HB donors (the two H atoms) and two acceptors (the two lone pairs of electrons in the outer orbital of O). Finding alternative, small molecules able to form a network of HBs, is not easy. Ammonia, with three HB donors and one acceptor (in the outer orbital of the N atom), can form a simpler network. Methane is not appropriate, having only donors, and no acceptors. Formamide, with three donors and three acceptors, has a high capability of hydrogen bonding, but has a relatively large size. Another key property of water molecules is that they are directly involved in the synthesis (dehydration) and breakdown (hydrolysis) of terrestrial biomolecules. Finding a medium with similar capabilities restricts the choice of potential alternatives. The possibility of a biochemistry not based on water should be tested with dedicated experiments.

1.2.1.2. *Properties of genetic and catalytic molecules*

Genetic molecules need long atomic chains to store a large number of units of information. Catalytic molecules need specific 3D conformations, which in turn require the folding of long atomic chains[4] with the aid of intramolecular forces. Whatever topology the long atomic chains of catalytic and genetic molecules may have (linear, branched, etc.), the backbones of these molecules must be built with atoms that have the ability to form structures with a variety of conformations. Owing to its electronic structure, carbon is the only atom of the periodic table that has the flexibility to form 1D, 2D and 3D molecular structures, including aromatic rings. *We therefore expect genetic and catalytic molecules to be long and mostly composed of carbon atoms*, not only in terrestrial life, but also in any form of life.

---

3 The hydrophobic effect may arise in any medium with a network of HBs (not only water).
4 Short atomic chains may have very limited catalytic properties, if any.

As we have seen, genetic and catalytic molecules should be able to interact, at least in part, via HBs (section 1.1.1.3). Therefore, they should have functional groups with hydrogen-bond donors and/or acceptors. This restricts the choice of suitable elements, because only some elements of the periodic table are more electronegative than H and can form donors X–H (see Note 1.1). If we restrict our attention to cosmically abundant elements – for example, those with abundance by number $X/H > 10^{-6}$ – only C, N, O and S have this property. These elements can also play the role of hydrogen bond acceptors, having at least one lone pair of electrons in their outer orbitals.

1.2.1.3. Chemical elements

Based on the above discussion, we conclude that H, C, N, O, and S are required for hydrogen bond interactions, either in the active sites of catalytic/genetic molecules or in the molecules of the medium. Obviously, life will need more elements than these. For instance, in terrestrial life, P, K, and Na play a key role in tasks of energy exchange and charge transportation, and minor quantities of metals, such as Fe, are essential in catalytic molecules. Other elements, such as Si, may be employed for purely structural reasons (e.g. shells of diatoms). Which elements, in minor amounts, could be needed in hypothetical non-terrestrial biochemistries we cannot say. To cast light on the general requirements of life, it suffices to say that CHONS and a liquid molecular medium with hydrogen bond capabilities are essential.

1.2.2. Physical conditions

The temperature and pressure of active life should lie within the limits that allow its molecular medium to be liquid (section 1.1.1.2). For early life forms in direct contact with their environment, this requirement constrains the ambient physical conditions.

1.2.2.1. Pressure

The pressure of the triple point of a given substance, p_{tr}, determines the minimum value of pressure that allows the existence of the liquid phase. In the case of water, $p_{tr} = 0.61$ kPa. This is one of the lowest values of triple point pressure among substances that one might consider as a potential alternative to water. For instance, ammonia and methane have $p_{tr} = 6.1$ and 11.7 kPa, respectively (Rumble 2022). Therefore, $p_{tr} \simeq 0.61$ kPa is a safe lower limit of ambient pressure for life. Albeit rarely taken into account, this limit provides one of the tightest constraints for the existence of life-supporting environments, because the vast majority of the volume of the universe is occupied by diffuse gas with extremely low pressure, such as the intergalactic medium, the interstellar medium or the interplanetary medium. Even in molecular clouds with $\approx 10^3$ atoms cm^{-3}, the densest of these diffuse media, the pressure is $\lesssim 10^{-10}$ Pa, many orders of magnitude below 0.61 kPa.

A loose upper limit for pressure can be established by assuming (as we do) that life processes are of chemical type. If the pressure is exceedingly high, the interatomic distances will be in the range of nuclear interactions ($\approx 10^{-15}$ m), rather than in the range of chemical interactions ($\approx 10^{-10}$ m). In many compact objects, such as neutron stars, life cannot exist, if nothing else, for this reason.

1.2.2.2. Temperature

Triple point temperatures, T_{tr}, provide a lower limit on the temperature of the liquid phase for any value of pressure[5]. The triple points of hypothetical alternatives to water, that is, cosmically abundant molecules with hydrogen bond capabilities (section 1.2.1.1), such as NH_3, HCN, H_2S, CH_3NO, CH_2O_2, yield a limit $T_{tr} \gtrsim 1.9 \times 10^2$ K. This limit precludes the habitability of low-temperature regions of the Universe, even though life with suspended metabolism and reproduction may exist in such regions.

The internal temperature of life processes is limited by the thermal impact of the molecular medium on the genetic and catalytic molecules. The thermal energy, $\varepsilon(T)$, is constantly exchanged via molecular collisions and, to avoid collisional destabilization of these molecules, should be $\varepsilon(T) < \Delta H_s$, where ΔH_s is the weakest dissociation energy of the intramolecular bonds that keep their 3D structure. If catalytic or genetic molecules lose their structure, they lose their functionality (they are said to be denaturated). The upper thermal limit of terrestrial organisms with active metabolism and reproduction, $\simeq 3.8 \times 10^2$ K (Clarke 2014), is likely due to the onset of denaturation. An independent upper limit of the same order can be obtained by assuming that the dissociation energies of the intramolecular bonds, ΔH_s, are in the range typical of hydrogen bonds (Vladilo and Hassanali 2018). Given the possible range of dissociation energies of HBs, the limit derived in this way is approximate. Even so, this limit excludes the habitability of any type of star because, whatever the star mass or evolutionary status, the temperature of the outer stellar layers always exceeds a few 10^3 K. The situation is obviously worse in stellar interiors, where the temperature increases from the outer layers to the thermonuclear core.

---

5 In general, the minimum temperature of the liquid phase of a given compound increases with pressure; water is a well-known exception, but its freezing point does not change significantly with pressure.

1.2.2.3. Protection from ionizing radiation

The Universe is filled with particles and photons, generated by stellar explosions, that are sufficiently energetic to ionize atoms and molecules. High doses of ionizing events can affect the structure of catalytic and genetic molecules, either directly or indirectly (via formation of free radicals in the molecular medium). The biological damage depends on the rate and type of ionizing particles and on the type of organism (Magill and Galy 2005). Even if there are no clearcut thresholds for radiation damage, and very low doses of radiation may even be beneficial for life, it is clear that life-nurturing environments must feature some form of protection from the intense ionizing radiation of cosmic origin.

1.2.3. Habitable worlds

The physical limits discussed above imply that stars and diffuse media must be discarded from the list of habitable environments. So, proceeding by exclusion, the only other astronomical environments that may host life are gravitationally accreted bodies not powered by thermonuclear reactions. In practice, these can be planets, dwarf planets, small bodies and moons[6]. Giant planets are not habitable, because their extended gaseous envelopes are characterized by strong atmospheric circulation and large gradients of pressure and temperature. Even assuming that a giant planet features atmospheric layers with habitable conditions (see Sagan and Salpeter 1976, for the case of Jupiter), hypothetical forms of life floating in such layers would be transported to non-habitable layers by atmospheric circulation. At variance with gaseous giants, rocky astronomical bodies (preferentially planets or moons) can offer surface or sub-surface layers with stable conditions of habitability, as long as their stellar, orbital and planetary properties are properly fine-tuned (see below). As far as we know, these rocky worlds are the only environments potentially habitable in the Universe.

1.2.3.1. Pressure and ionizing radiation

The surface of a rocky planet (or moon) can satisfy the minimum pressure requirement (section 1.2.2.1) if it is covered by an atmosphere that yields a surface pressure $p_s \gtrsim 0.61$ kPa. The atmosphere also protects the surface from ionizing radiation, since the surface dose of cosmic radiation decreases with increasing mass

---

[6] These bodies are classified according to their ability to clear their orbits from debris or accreting sufficient mass to attain a spherical shape. Planets and dwarf planets attain spherical shape, but only planets clear their orbits; small bodies, such as asteroids and comets, do not attain spherical shape and do not clear their orbits. Moons orbit one of these bodies; they may or may not attain spherical shape.

of the atmospheric column (Atri et al. 2013). Atmospheres can be present in planets or moons with mass sufficiently high to gravitationally capture volatile compounds during their formation; smalls bodies, such as solar system asteroids and comets, are unlikely to satisfy these conditions. In bodies without atmospheres, the minimum pressure requirement and protection from cosmic radiation might be achieved at some depth below the surface.

1.2.3.2. Climate and temperature

The surface temperature of a planet is determined by its climate, which in turn is regulated by the energy balance between incoming stellar radiation and outgoing long-wavelength radiation (Pierrehumbert 2010). The global incoming radiation is given by $S(1-A)$, where S is the instellation, that is, the radiative flux received by the star, and A the albedo, that is, the fraction of stellar photons reflected or scattered back in space without heating the surface. The instellation scales with the star luminosity, $L_\star$, and the planet-star distance, d, according to the law $S = L_\star/(4 \pi d^2)$. The albedo depends on the reflectivity of the surface and the scattering properties of the atmosphere. The out-going radiation varies with the planet temperature, but is controlled by the greenhouse effect, which depends on the structure and chemical composition of the atmosphere. The planetary energy balance is also influenced by clouds through their contribution to the albedo and the greenhouse effect. Depending on all the above-mentioned factors, rocky planets or moons may have surface temperatures ranging from just above 0, up to a few 10^3 K. In this range, optimal conditions of habitability can be achieved with a suitable combination of stellar, orbital and planetary properties.

1.3. Setting the stage for chemistry and life in the Universe

1.3.1. *Births of the laws of chemistry*

The chemical history of our Universe that ultimately led to the origin of life and formation of habitable planets started long time ago. Similar chemical histories may have taken place in different regions of the Universe, and not just where we live. However, we might wonder whether or not the familiar laws of chemistry are valid at any epoch and in any region of the Universe. The physical properties of atoms and molecules – and hence their chemical interactions – are governed by two dimensionless numbers: the fine-structure constant, $\alpha = e^2/(h/2\pi) c = 7.2972 \times 10^{-3}$, and the electron-to-proton mass ratio, $\beta = m_e/m_p = 1836^{-1}$ (see Barrow 2022). Should α and β have different values, the laws of chemistry would be very different to the point that life would not be able to emerge[7]. The possibility that α and β may vary in space and time has been advanced in studies of theoretical physics, but, so far, observational studies have not detected relative variations of α and β over cosmological distances, within experimental errors of a few ppm (Martins 2017). Therefore, we can apply the familiar laws of chemistry to study prebiotic or planetary-formation chemical pathways in different epochs and regions of the Universe, such as distant molecular clouds, the primitive Earth, the young Milky Way and the young Universe.

1.3.2. Production of chemical elements

Our understanding of the early stages of the Universe relies on hot Big Bang models, which are supported and constrained by a wealth of high-energy experimental data and astrophysical observations (Cyburt et al. 2016). Starting at ≈ 1 sec after the initial singularity, the physical conditions of the Universe allowed free protons and neutrons to form a few atomic nuclei (D, ^{3}He, ^{4}He, ^{7}Li) by nuclear fusion. A few minutes later, the expanding universe cooled down to a point where nucleosynthesis of heavier elements was not possible. Therefore, none of the CNO atomic nuclei, or heavier elements useful for life, were synthesized during the Big Bang.

After the initial nucleosynthesis, the Universe was largely composed of free protons and electrons, photons and neutrinos. The interactions were still too energetic to allow for the formation of neutral atoms. Later on, after $\sim 3.7 \times 10^5$ year, the temperature dropped to $\simeq 3,000$ K and neutral H could form from radiative recombination of protons and electrons. The recombination generated a flash of ultraviolet photons which, as a result of the cosmological expansion of the Universe, are presently seen as a diffuse microwave radiation: the cosmic microwave background (CMB). Detailed analysis of the CMB shows that at the epoch of recombination the matter distribution was relatively smooth with temperature fluctuations as small as $\delta T/T_0 \simeq 4 \times 10^{-5}$. The growth and gravitational collapse of overdensity regions lead to the formation of the first stars and quasars, which gradually reionized the gas. According to model predictions, the stars of the very first stellar population (called Pop III) were able to produce and eject carbon and heavier elements for the first time (Hirano et al. 2014). This marks the beginning of carbon-based chemistry in the interstellar space.

---

7 This argument is an example of the so-called anthropic principle (see Barrow 2002).

About 1 Gy after the Big Bang, the first galaxies formed from the collapse of large-structure overdensity regions. The subsequent history of chemical evolution in our galaxy can be traced back thanks to the huge amount of spectroscopic data of Milky Way stars in conjunction with models of stellar nucleosynthesis and stellar explosions (Maiolino and Mannucci 2019). The long-lived, low-mass stars of the old stellar population, called Pop II, are still visible and are characterized by a spherical distribution and a very low metallicity[8]. The most recent stellar population, called Pop I, which has a disk distribution, is concentrated in the spiral arms of our galaxy, and has a higher metallicity, close to the solar value. This chemical enrichment is the result of the ejection of stellar products to the interstellar gas via stellar explosions and the subsequent incorporation of enriched interstellar gas in stars of successive generations (Nomoto et al. 2013).

The present-day, average cosmic composition resulting from this complex history shows that the elements most important for life, such as H, C, N and O, are particularly abundant. The same is true for Si and Fe, required to form habitable, rocky planets. To understand if the products of astrophysical nucleosynthesis can be effectively incorporated in life-hosting environments or life, we must track the evolution of chemical compounds from interstellar clouds to planetary systems (see Ziurys et al. (2016), for example, for the history of carbon).

1.3.3. *Assemblage of prebiotic molecules*

1.3.3.1. *Prebiotic chemistry in space*

More than 200 molecular species have been detected in space, either in the inter-stellar medium (ISM), in the vicinity of star forming regions, or in the circumstellar medium, especially around evolved stars (McGuire 2018). The interstellar molecules containing at least six atoms are carbon-based and are called COMs (complex organic molecules) in the astronomical community, despite their small size for biological standards. COMs are synthesized in dense molecular clouds, which are the cradles of stars and planetary systems. Several COMs are remarkably interesting in the context of prebiotic chemistry. Examples include glycolaldehyde (the simplest sugar), aminoacetonitrile (a possible precursor of glycine) and even potential RNA precursors (Jiménez-Serra et al. 2020). Also formamide, a key prebiotic precursor, has been detected in the ISM (López-Sepulcre

---

8 In astrophysics the metallicity is defined as [X/H] = log (X/H) − log (X/H)$_{sun}$, where X is an element at least as heavy as C (often Fe), (X/H) is the abundance ratio by number measured in an astronomical source, and (X/H)$_{sun}$ is the same ratio measured in the Sun.

et al. 2019). The existence of prebiotic molecules in space discloses the possibility that, at the stage of star and planetary formation, part of these molecules may be incorporated in planetary system bodies.

1.3.3.2. Prebiotic chemistry in star forming regions

A large amount of chemical processing takes place during star formation (Jørgensen et al. 2020). The main stages of formation of solar-type stars, the most interesting for habitability (section1.4.1.1), can be summarized as follows (Caselli and Ceccarelli 2012). First, a dense prestellar core forms inside a giant molecular cloud. Matter accumulates toward the center, where atoms and molecules in the gas phase freeze-out onto the surfaces of the dust grains, forming icy mantles. The mobility of H atoms on the grain surfaces allows hydrogenated species to be formed (e.g. water, formaldehyde, methanol and so on). In a second stage of collapse, a protostellar envelope is formed around a hydrostatic core. The envelope warms up due to the conversion of gravitational energy into radiation. When the temperature rises above $\simeq 100$ K, the icy mantles sublimate, giving rise to complex organics in the so-called hot corino regions. During the collapse, a fraction of matter is ejected, forming supersonic jets and molecular outflows. The shocks between this material and the quiescent gas of the envelope evaporate part of the grain mantles and refractory grains. In a third stage, the protostar is formed and the envelope gradually dissipates, leaving a rotating protoplanetary disk with a typical size[9] in the range $\approx 10^2-10^3$ AU. In the inner, hot regions of the disk, new complex molecules are synthesized, while in the outer, cold regions the molecules inherited from the protostellar phase freeze-out again onto the grain mantles. The degree by which the chemistry of the parent cloud is altered as material is accreted to the disk is still unclear.

1.3.3.3. Prebiotic chemistry in planetesimals

The gradual assemblage of part of the gaseous and solid components of the protoplanetary disk gives rise to small bodies called planetesimals (Blum and Warm 2008). Planetesimals not accreted by planets may preserve a pristine record of the organic material synthesized in the parent cloud. In the solar system, the existence of this connection can be tested by analyzing asteroids and comets, which are examples of pristine planetesimals. Large organic inventories have been found from in situ research of cometary chemistry (Altwegg et al. 2019). Analysis of these findings makes it plausible that this organic material was accreted almost unaltered from the prestellar stage.

---

9 The mean distance between the Sun and the Earth, called Astronomical Unit (AU), is used to express distances inside planetary systems and protoplanetary disks.

Asteroid chemistry can be probed from the analysis of fragments collected on Earth in the form of meteorites. Carbonaceous chondrite (CC) meteorites are a group of such fragments containing organic materials with structures as diverse as macromolecular insoluble organic material (IOM) and simple, soluble compounds of prebiotic interest, such as amino acids (Pizzarello and Shock 2017). Several organic compounds found in CC meteorites are rich in the heavy isotopes of C, N and H. The measured isotopic ratios suggest that part of this organic material originated in the parent cloud (Pizzarello and Shock 2017).

In addition to amino acids, guanine, adenine and uracil have been found in several CC meteorites, indicating that the nucleobases of terrestrial life can be formed in asteroids (Pearce and Pudritz 2016). The findings of key prebiotic molecules in planetesimals make the plausible hypothesis that prebiotic compounds originated in space and subsequently delivered to Earth may have contributed to the origin of terrestrial life (see section 1.5.1).

1.3.4. Origin of water

The chemical history of water is a journey from interstellar clouds to protoplanetary disks, planetesimals and planets (van Dishoeck et al. 2021; Ceccarelli and Du 2022). Water is initially synthezised in dense molecular clouds by the hydrogenation of frozen oxygen on the surface of interstellar dust grains. During protostellar collapse, water stays mostly in ice mantles, with the exception of the inner regions (the hot corinos), where water vapor is released due to ice sublimation.

Water is transported from clouds to protoplanetary disks mostly as ice, even though water vapor reservoirs are also present in disks. The inner boundary of the region of the protoplanetary disk where ice condenses is called the snow line. The location of the snow line affects planet formation, because ice mantles on dust grains promote the growth of solids to planetesimal sizes. The equivalent amount of many Earth oceans of water are likely present in the planet forming regions of disks. However, it is unknown which fraction of this water is effectively incorporated in planets formed inside the snow line, because planetesimals originating inside the snow line are expected to be dry. After planetary formation, volatile-rich planetesimals that have been formed beyond the snow line may migrate due to dynamical instabilities and may deliver water to rocky planets formed inside the snow line. In the case of the Earth, this scenario of late delivery of water is supported by the similarity of the D/H isotopic ratio measured in Earth's oceans and in water-rich asteroids.

1.3.5. Appearance of rocky planets

Chemistry plays an essential role not only for the assemblage of prebiotic compounds, but also in the early stages of the formation of potentially habitable worlds, that is, rocky planets or moons (section 1.2.3). The formation of rocky planets starts from the sticking/accretion of refractory dust particles, that is, grains rich in silicates, in protoplanetary disks. The solid component undergoes several steps of accretion until planetesimals, with sizes in the order of several kilometers, are formed (Blum and Wurm 2008). Collisions and gravitational interactions between planetesimals generate planetary embryos, with sizes in the order of 10^3 km. Eventually, the embryos accrete planetesimals and collide until they form rocky planets. Planetary formation is very complex, because these processes largely overlap in time, are driven by different physical mechanisms and span over 12 orders of magnitude in size of the accreting solids.

The initial efficiency of the accretion of solids depends on the amount of dust which, in turn, depends on the abundance of metals inherited from the parent cloud. Planet formation models usually assume a present-day solar chemical composition and a standard value of interstellar dust-to-gas ratio by mass ($\simeq 10^{-2}$). Since the dust-to-gas ratio scales with the metallicity of the gas, the amount of dust is expected to be smaller in astronomical regions of lower metallicity. As a result, the formation of rocky planets may be inhibited below some threshold of metallicity. Despite the lack of direct observational evidence, it is probably safe to say that planet formation did not take place around the very first generation of stars (Pop III stars), which were born from gas free of carbon or heavier elements. It is instead possible that planets may have been assembled around Pop II stars, which were formed in the young Milky Way from gas with a modest, but not negligible, amount of metals.

So far, attempts to detect planets around Pop II stars have not been successful (Wallace et al. 2020; Yoshida et al. 2022). We know that giant planets become less frequent with decreasing metallicity of the host star, but this trend may not apply to rocky planets (Zhu and Dong 2021). If planets around Pop II stars are predominantly rocky, their detection is challenging because rocky planets are small and Pop II stars are faint. Future observations will clarify whether or not planets had already appeared in the young Milky Way. At the moment, we know that planets are quite common around the most recent stellar population (Pop I) that was formed from clouds enriched by metals in the course of Galactic chemical evolution (section 1.3.2). The findings of statistical numbers of planets orbiting stars other than the Sun (exoplanets) have boosted the search for life-hosting environments outside the solar system.

1.4. The habitable Universe

For about four decades after the Miller experiment, the search for life outside of Earth was restricted to solar system planets or moons. This situation has completely changed after the discoveries of the first exoplanets (Wolszczan and Frail 1992; Mayor and Queloz 1995). In the course of the past three decades, a large number (currently $\simeq 5.3 \times 10^3$) of exoplanets have been detected using direct and indirect observational techniques (Perryman 2018). Whatever technique is employed, rocky planets are more difficult to detect than giant planets because of their relatively small mass, radius and emissivity. However, when observational biases are taken into account, a remarkable result is found: small planets are intrinsically numerous. In the inner region of planetary systems ($\lesssim 1$ AU), about 30% of Sun-like stars host planets with masses and/or radii down to Earth mass and/or radius, and each planetary system has, on average, about three such planets (Zhu and Dong 2021). Besides Earth-like planets, exoplanet surveys have also discovered slightly bigger planets, called Super-Earths, that are likely to be rocky, and so potentially habitable, as long as their radius is sufficiently small ($r \lesssim 1.9$ R_{earth}). Planets with larger radii are believed to accrete a large amount of volatiles during planetary formation, ending up as planets with extended (non-habitable) atmospheres.

The abundance of rocky planets indicates that many habitable environments may exist, in spite of the need for fine-tuning many stellar, orbital and planetary properties for a planet to be habitable (section 1.2.3). Aside from these statistical results, the time is ripe to start assessing the habitability of individual exoplanets.

Characterizing a specific exoplanet is not easy, due to the limited amount of observational data available for each individual case. The instellation (section 1.2.3) and its seasonal variation can be estimated from the luminosity of the host star and the orbital parameters. The planetary mass, m, and radius, r, can be measured with the radial-velocity and transit method, respectively (Perryman 2018). Only in a small number of cases (currently $\simeq 2 \times 10^2$ with $r \leq 2$ R_{earth}), both techniques can be applied, allowing the determination of the surface gravity acceleration, $g \propto m/r^2$, mean density, $\rho \propto m/r^3$ and escape velocity, $v_{esc} \propto (m/r)^{1/2}$. With the aid of models of planetary structure, these data allow us to discriminate rocky planets from other types (e.g. gaseous, water-dominated and so on). Apart from these and some other data, a vast amount of planetary properties that govern the climate and habitability (e.g. rotation period, axis tilt, atmospheric structure and composition, land/ocean distribution, etc.) are unconstrained by observations. To assess a possible range of habitable conditions of a specific exoplanet, we can treat these unknown properties as free parameters in specifically designed, fast climate models (see, for example, Silva et al. (2017b), for the case of Kepler 452b). However, a proper characterization of individual rocky planets must await direct measurements of their atmospheric properties (Wordsworth and Kreidberg 2022), which will provide tight constraints on the climate and habitability.

1.4.1. Circumstellar habitable zones

The fact that the instellation scales as $S \propto d^{-2}$ (see section 1.2.3.2) has lead to the concept of (circumstellar) HZ, a range of star-planet distances, d, where the planet surface can keep water in the liquid phase over geological time scales. The first calculations of the inner and outer edge of the HZ were performed in two extreme cases: (i) a hot, water-rich atmosphere and (ii) a CO_2-dominated atmosphere, respectively (Kasting et al. 1993; Kopparapu et al. 2013). The inner and outer edges were calculated for different types of main-sequence stars, that is, stars in the most stable stage of hydrogen nuclear burning. The results indicate that the Earth lies inside the HZ of a solar-type star, as it should, but Venus and Mars lie just outside the inner and outer edges, respectively. Planets inside the inner edge, like Venus, are predicted to undergo a runaway greenhouse instability, leading to the full loss of their water reservoirs (Kasting et al. 1993). With a proper atmospheric composition, planets may be habitable also beyond the outer edge of the classic HZ. Examples include planets with hydrogen-rich or methane-rich atmospheres (Ramirez and Kaltenegger 2017, 2018). Whether or not life can actually thrive in atmospheres with extremely non-terrestrial compositions should be investigated.

A central concept of the HZ is the existence of a geochemical cycle that allows the long-term stabilization of the climate through a gradual change of atmospheric composition and greenhouse effect. The stabilization mechanism assumed to be at work is the cycle of inorganic CO_2, a gas that can be removed from or injected into the atmosphere, depending on the surface/atmospheric temperature (Kasting and Catling 2003). If this cycle is essential for the climate stabilization, only planets with geophysical activity, including volcanoes and plate tectonics, may support long-term habitability. Not accounting for this stabilizing cycle yields narrower HZs (e.g. Silva et al. 2017a). Also biological cycles affect the long-term habitability, because life can gradually change the composition of the overlying atmosphere (see Figure 1.1) and the reflectance of the planetary surface. The resulting changes of greenhouse effect and planetary albedo may generate stabilizing or destabilizing climate effects.

Most studies of the HZ are based on results obtained from 1D (single atmospheric column) climate models. More refined calculations should be performed with 3D, global circulation models (GCMs) (see Pierrehumbert 2010). Unfortunately, this approach has practical limitations, because GCMs are too time consuming for a systematic exploration of the huge parameter space of non-terrestrial climate factors. A preliminary exploration of this type can be performed using climate models of low or intermediate complexity (see, for example, Biasiotti et al. (2022); Paradise et al. (2022)).

1.4.1.1. Host star properties and circumstellar habitability

The instellation $S = L_\star/(4\pi d^2)$ is a strong function of the mass of the star, $M_\star$, because the luminosity of main-sequence stars scales as $L_\star \propto M_\star^{3.5}$. The very low luminosity of stars with $M_\star \lesssim 0.5\ M_{sun}$ implies that the distance between the HZ and these stars is very small ($d \lesssim 0.1$ AU). The effective habitability of planets in these close-in HZs is uncertain because (i) low-mass stars have strong stellar activity and their energetic events may disrupt the atmospheres of close-in planets; (ii) planet-star tidal interactions are so strong that the orbital and rotational rotation periods may become synchronized, with one planet hemisphere constantly illuminated and the other constantly dark; (iii) planets in the main-sequence HZ may have undergone a runaway greenhouse instability during the pre-main-sequence stage, which is characterized by a significantly higher stellar luminosity (Ramirez and Kaltenegger 2014).

The variation of star luminosity in the course of stellar evolution shifts the location of the HZ. During the main-sequence stage, the luminosity rises gradually and the planet may maintain a moderate climate with the aid of a geochemical/atmospheric mechanism of stabilization. At the end of this stage, however, the star undergoes a strong and fast rise of luminosity that shifts the HZ outwards. At this point, the planet cannot compensate the corresponding rise of S and undergoes a runaway greenhouse instability. The main-sequence stage lasts several Gyr for low-mass stars but can be as short as few 10^8 year for stars with masses $\gtrsim 2\ M_{sun}$. The latter time scales are much shorter than those of terrestrial biological evolution. Therefore, planets around massive stars are not able to host life for long geological periods and, in the most extreme cases, they may not have enough time for the emergence of life.

The HZs around stars with solar-type masses are sufficiently distant from the star to avoid the problems of close-in HZs. At the same time, the HZs around solar-type stars can benefit from a stable instellation for several Gyrs. Therefore, planets in the HZs of solar-type stars are the best candidates for long-term habitability. Unfortunately, the presence of selection effects makes more difficult to detect planets in the HZ of solar-type stars than in the HZ of low-mass stars (e.g. Perryman 2018). For this reason, optimal candidates for planetary habitability are still rare, but this situation will change with increasing statistics of exoplanets.

The HZ is usually defined for planets orbiting a single star, but a large fraction of Milky Way stars (≈50%) belong to binary stellar systems. The habitability in binary systems is influenced by the total instellation and the gravitational perturbations generated by the two stars. Dynamically stable orbits exist if the planet is sufficiently close to one of the two stars (S-type orbits), or sufficiently distant from both stars (P-type orbits). Statistical studies indicate that the instellation of P-type orbits is generally too low for habitability, whereas S-type orbits offer, on average, an instellation suitable for habitability (Simonetti et al. 2020).

1.4.2. Galactic habitable zones

The galactic habitable zone (GHZ) was defined as an annular region lying in the plane of the Milky Way disk possessing the heavy elements necessary to form rocky planets and long-term clement conditions to allow for the evolution of multicellular life (Gonzalez et al. 2001; Lineweaver et al. 2004). The location and extension of the GHZ are calculated by convolving (i) the probability of rocky planet formation, estimated as a function of the metallicity, and (ii) the probability of life destruction, estimated from the rate of supernova explosions. Both probabilities are calculated as a function of galactocentric radius, r_{gc}, and time, t, with the aid of models of galactic chemical evolution. The results indicate that the GHZ lies in the range $7 \lesssim r_{gc}$ (kpc) $\lesssim 9$ and widens in time (for comparison, the Sun lies at $r_{gc} \simeq 8.1$ kpc). Calculations of the GHZ have been extended to other spiral galaxies, using upgraded recipes of galactic chemical evolution: for M31 (Andromeda), the maximum of habitability is found at $r_{gc} \simeq 16$ kpc (Spitoni et al. 2014). Calculations performed with high-resolution cosmological simulations of galaxy formation allow the mapping of the 3D structure and temporal evolution of the GHZ (Forgan et al. 2017).

The definition of the GHZ should be taken with caution because (i) the relationship between metallicity and probability of rocky planet formation is still not clear (Zhu and Dong 2021); (ii) rather than sterilizing a planet, supernovae may alter the biological evolution, leading to the disappearance of some species and the formation of new ones (Prantzos 2008). An approach to tackle this latter problem is to assume that the evolution is reset (e.g. restarting from unicellular life) at each critical SN event (Morrison and Gowanlock 2015). Recent work has extended GHZ calculations to Milky Way globular clusters (Di Stefano and Ray 2016), dwarf galaxies (Stojković et al. 2019) and elliptical galaxies (Lacki 2021). However, the possibility of rocky-planet formation and long-term habitability in low-metallicity, dense stellar regions, such as globular clusters, is currently unknown.

1.5. Planetary environments suitable for the origin of life

As long as the requirements for habitability are satisfied (see section 1.2), life can colonize any planetary environment from the surface, to the atmosphere, to the deep ocean. However, the conditions for habitability do not imply that life can emerge. In fact, the conditions for abiogenesis could be different from, and perhaps tighter than, the conditions for habitability (Chopra and Lineweaver 2016; Cockell et al. 2016). To find out if life may effectively exist in a habitable environment, we should try to understand which ambient conditions are suitable, or unsuitable, for abiogenesis.

Starting from Miller (1953), a large number of laboratory experiments have succeeded to assemble, one at a time, the basic molecular ingredients of terrestrial life, such as amino acids, nucleobases and compartments (e.g. Luisi 2016; Saladino et al. 2019 and references therein). A general conclusion that can be inferred from these experiments is that, in order to be successful, each step of prebiotic chemistry requires its own specific ambient conditions. In other words, life cannot emerge in a single pot with steady conditions. This inference suggests that the planetary environment where life may potentially emerge should provide (i) the possibility of transporting chemicals between sites with different physicochemical conditions and/or (ii) sites that undergo suitable changes of physicochemical conditions. In the rest of this section, these (and other) requirements are used to contrast two environments of abiogenesis proposed in the literature: the planetary surface, including shallow waters (Toner and Catling 2019; Damer and Deamer 2020 and references therein) and the bottom of the oceans (Baross and Hoffman 1985; Westall et al. 2018 and references therein).

1.5.1. Abiogenesis on planetary surfaces

Planetary surfaces can offer vastly diverse physicochemical conditions in subaerial sites (ponds, lakes, beaches, dry areas), especially if geophysical activity is present (volcanoes, geysers, hot springs). If needed, the transfer of prebiotic products between different sites can be carried out with the aid of a hydrological system. Planetary surfaces also offer sites with cyclic variations of physicochemical conditions, such as changes of stellar irradiation, provided by the day-night cycle, and changes of concentrations, induced by evaporation and condensation of shallow waters (Toner and Catling 2019), or by tidal cycles on beaches (Bywater and Conde-Frieboesk 2005). The possibility of varying the concentration determines the success or failure of any chemical reaction, including the prebiotic ones. Specific mechanisms for enhancing concentrations in surface lakes may help provide the phosphate, which is required for the origin of terrestrial life (Toner and Catling 2020). Also, the exposure to stellar radiation can be fine-tuned: for instance, attenuation of ultraviolet radiation, if needed, can be obtained at very moderate depths below the surface (e.g. a few millimeters in solid surface or a few centimeters in water). In addition, rocky surfaces offer minerals and metals that can provide replication templates or catalytic properties invoked in some scenarios of abiogenesis.

Dynamical instabilities leading to collisions and ejection of fragments of planetesimals were likely frequent at the epoch of the origin of life. Planetary surfaces can collect prebiotic chemicals originating in space, incorporated in planetesimals (section 1.3.3.3), and delivered on Earth (Cronin and Pizzarello 1997; Pizzarello and Williams 2012). A problem with the hypothesis of exogenous delivery is that the meteors melt by overheating while traveling at high speed through the Earth's atmosphere. However, when meteorites are large enough, after ablation of the outermost layers during fall, the remaining heated layers solidify into a thin fusion crust that leaves the cold interior untouched and still cold (Pizzarello and Shock 2017). The most energetic impacts may have destructive effects, but they may also provide alternative pathways of prebiotic synthesis on planetary surfaces (Ferus et al. 2020). The interaction of the surface with an impact-generated atmosphere rich in CH_4, H_2 and NH_3 offers an additional possibility for the synthesis of organic compounds on the early Earth (Zahnle et al. 2020). This possibility is extremely important, because reduced carbon and nitrogen increase the probability of success of prebiotic chemical pathways.

1.5.1.1. *Origin of molecular chirality*

The origin of the chirality of biological molecules (see Note 1.2) is still an open question in studies of the origin of life (see Luisi 2016). The general idea is that, after an initial, small enantiomeric excess (*ee*), some form of chiral amplification (Bryliakov 2020) allowed homochiral polymers to be assembled. Amino acids produced in Miller-type experiments do not show any trace of *ee*, but, remarkably, several amino acids found in CC meteorites contain an *ee* of L-type, the same type of biological amino acids (Pizzarello 2016). The presence of an *ee* in CC meteorites offers an unequivocal indication that a purely chemical evolutionary process is able to generate chiral selection. Whether a small *ee* brought by meteorites in the primitive Earth could have initiated the homochirality of terrestrial biomolecules, we cannot say. However, to avoid severe dilution of the exogenous molecules, this hypothesis is plausible only if the organic material is collected by a solid surface, rather than an open ocean.

An alternative, non-exogenous hypothesis is that an initial chiral excess was generated in the Archean Earth. Among scenarios of this type recently proposed, an ee could have been induced by spin-polarized electrons produced by UV irradiation of magnetite deposits (Ozturk and Sasselov 2022) or by β^- decays of ^{40}K in potassium-rich prebiotic solutions (Vladilo 2022). Both scenarios require the presence of modest amounts of water (ponds or shallow lakes) on a solid planetary surface. Also in these non-exogenous scenarios, an open ocean would not work.

NOTE 1.2.– Molecules with identical chemical formula and chemical bonds are said to be chiral when they are nonsuperimposable mirror images of each other. The two mirror images of a chiral molecule are called enantiomers. The amino acids and sugars of terrestrial life are chiral and appear in only one enantiomeric form. For instance, protein amino acids only appear in the L-type chiral form.

1.5.2. Abiogenesis in the oceans

Hydrothermal vents at the bottom of the oceans have been proposed as possible sites for a submarine origin of terrestrial life (Baross and Hoffman 1985). The cold ocean water percolates into the oceanic crust, is heated by an underlying, geophysically active area, and reemerges to form the vents. In the hydrothermal vent scenario, the physical and chemical gradients associated with the vents are assumed to play a key role in the origin of life. In addition to vents over hot magma, other sites have been proposed, including alkaline vents with moderate temperatures (Sojo et al. 2016) and the sedimentary layer between oceanic crust and seawater (Westall et al. 2018). The hydrothermal vent hypothesis postulates an autotrophic origin of life, with organic compounds directly synthesized by geochemical reduction of CO_2 by H_2.

A general problem with the vent scenario is that the physicochemical conditions of submarine sites are rather steady. Since each stage of abiogenesis requires its own specific conditions, this is a serious disadvantage. More specifically, the submarine scenario has the following drawbacks (see Deamer et al. 2019): (i) assuming that small organic compounds are formed, these products would be diluted into the ocean, preventing the possibility of their concentration for the synthesis of prebiotic polymers; (ii) the condensation of amino acids to peptides, or nucleotides to nucleic acids is inhibited in the presence of water; (iii) the chemicals required for essential prebiotic steps, such as cyanide or phosphate, would be too diluted in the ocean. Another problem is the formation of membranes of protocells, which requires reservoirs dominated by K and Mg salts, while terrestrial oceans are dominated by Na salts (Natochin 2010). Moreover, as explained above, understanding the origin of molecular chirality is even more problematic in the oceans than on the surface. Finally, it is true that submarine sites offer protection from exogenous impacts but, on the other hand, they cannot exploit the prebiotic potential of exogenous delivery.

1.5.3. Implications for the search for life outside Earth

The autotrophic scenario of abiogenesis in the oceans presents problems which seem too difficult to circumvent. Instead, planetary surfaces offer many different possibilities to process organic compounds assembled in situ or delivered from space, in the framework of a heterotrophic origin of life (Lazcano 2016). Even if the scenarios considered are terrestrial, many of the above arguments are general and the conclusions can be extended to non-terrestrial worlds. In the solar system, the drawbacks of the submarine scenario of abiogenesis cast doubts on the possible presence of life in the oceans below the surface of the icy moons of Jupiter (see also Pascal 2016). Beyond the solar system, exoplanet surveys have discovered many planets covered by global oceans (Kite and Ford 2018). Even if many of these ocean worlds lie in the HZ of their host stars, the difficulties of a submarine abiogenesis make the presence of life on such planets quite doubtful. The existence of sites suitable for the origin of life should be taken into account when searching for inhabited worlds outside of Earth.

1.6. The quest for inhabited worlds

Life spread on a planetary surface leaves a chemical imprint on the overlying atmosphere (Figure 1.1). The search for life on remote planets is focused on the possibility of detecting atmospheric compounds generated in this way, called atmospheric biosignatures (Catling et al. 2018). Atmospheric compounds can be detected by means of spectroscopic observations, but discriminating compounds of biotic and abiotic is quite a challenging task (Schwieterman et al. 2018). Since life creates a chemical disequilibrium (Figure 1.1), biotic products might be recognized by comparing abundances measured in the atmospheric spectra with abundances predicted by chemical-equilibrium models of atmospheric composition. Differences in the relative abundances of compounds of putative biological origin would reveal the presence of biological activity.

From the experimental point of view, exoplanet atmospheres can be probed with transmission, reflection and emission observational techniques. Transmission spectra can only be obtained for the exoplanets that transit in front of their host star. Reflection spectra can be obtained for planets closely illuminated by their host star when the orbital phase exposes the illuminated hemisphere to the observer. Both transmission and reflection techniques have a strong selection bias in favor of planets very close to the star. Therefore, these techniques can be applied to planets in the close-in HZs of low-mass stars, which may have severe problems of habitability (section 1.4.1.1), but not in the more detached HZs of solar-type stars, which are unaffected by such problems. Another limitation is that transmission spectroscopy is effective for planets with extended atmospheres, but not for habitable planets, characterized by thinner atmospheres.

Emission spectroscopy in the mid-infrared spectral window, where planetary emission is maximum, provides an alternative to transmission and reflection spectroscopy. Emission spectra of exoplanets can be obtained indirectly or directly. The indirect method is limited to transiting planets, since the spectrum is obtained by comparing observations performed outside and during a secondary transit, that is, when the planet lies behind its star. A more general and promising technique is the direct emission spectroscopy of the planet. This technique is very demanding because it requires space instrumentation with (i) interferometric capabilities to provide a spatial resolution sufficient to separate the planet from the star; (ii) capability of obscuring or canceling the emission of the star; (iii) large collecting area to detect the weak planetary signal; and (iv) instrumentation cooled down to a few kelvin to avoid thermal noise in the infrared spectrum. Projects with such ambitious goals (e.g. Barstow et al. 2021; Quanz et al. 2022) will need a few decades to be developed and, hopefully, implemented. If successful, they will give us the opportunity to detect signatures of life in planets orbiting different types of stars, including Earth-like planets around solar-type stars. A positive detection would prove that the cosmic chemical evolution that has led to the origin of terrestrial life has not been an isolated event. The same discovery would displace the Earth from the center of the biological world half a millennium after the Copernican revolution displaced our planet from the center of the astronomical universe.

1.7. References

Altwegg, K., Balsiger, H., Fuselier, S.A. (2019). Cometary chemistry and the origin of icy solar system bodies: The view after Rosetta. *Annual Review of Astronomy and Astrophysics*, 57, 113–155.

Atri, D., Hariharan, B., Grießmeier, J.-M. (2013). Galactic cosmic ray-induced radiation dose on terrestrial exoplanets. *Astrobiology*, 13, 910–919.

Baross, J.A. and Hoffman, S.E. (1985). Submarine hydrothermal vents and associated gradient environments as sites for the origin and evolution of life. *Origins of Life and Evolution of the Biosphere*, 15, 327–345.

Barrow, J.D. (2002). *The Constants of Nature: From Alpha to Omega.* Jonathan Cape, London.

Barstow, M.A., Aigrain, S., Barstow, J.K., Barthelemy, M., Biller, B., Bonanos, A., Buchhave, L., Casewell, S.L., Charbonnel, C., Charlot, R.S. et al. (2021). The search for living worlds and the connection to our cosmic origins. *Experimental Astronomy*, 54, 1275–1306.

Biasiotti, L., Simonetti, P., Vladilo, G., Silva, L., Murante, G., Ivanovski, S., Maris, M., Monai, S., Bisesi, E., von Hardenberg, J., Provenzale, A. (2022). EOS-ESTM: A flexible climate model for habitable exoplanets. *Monthly Notices of the Royal Astronomical Society*, 514, 5105–5125.

Blum, J. and Wurm, G. (2008). The growth mechanisms of macroscopic bodies in protoplanetary disks. *Ann. Rev. Astron. Astrophys.*, 46, 21–56.

Bryliakov, K.P. (2020). Chemical mechanisms of prebiotic chirality amplification. *Research*, 5689246.

Bywater, R.P. and Conde-Frieboesk, K. (2005). Did life begin on the beach? *Astrobiology*, 5, 568–574.

Caselli, P. and Ceccarelli, C. (2012). Our astrochemical heritage. *Astron. Astrophys. Rev.*, 20, 56.

Catling, D.C., Krissansen-Totton, J., Kiang, N.Y., Crisp, D., Robinson, T.D., DasSarma, S., Rushby, A.J., Del Genio, A., Bains, W., Domagal-Goldman, S. (2018). Exoplanet biosignatures: A framework for their assessment. *Astrobiology*, 18, 709–738.

Ceccarelli, C. and Du, F. (2022). We drink good 4.5-billion-year-old water. *Elements*, 18, 155–160

Chopra, A. and Lineweaver, C.H. (2016). The case for a Gaian bottleneck: The biology of habitability. *Astrobiology*, 16, 7–22.

Clarke, A. (2014). The thermal limits to life on Earth. *International Journal of Astrobiology*, 13, 141–154.

Cockell, C.S., Bush, T., Bryce, C., Direito, S., Fox-Powell, M., Harrison, J.P., Lammer, H., Landenmark, H., Martin-Torres, J., Nicholson, N. et al. (2016). Habitability: A review. *Astrobiology*, 16, 89–117.

Cyburt, R.H., Fields, B.D., Olive, K.A., Yeh, T.-H. (2016). Big bang nucleosynthesis: Present status. *Rev. Mod. Phys.*, 88, 015004.

Cronin, J.R. and Pizzarello, S. (1997). Enantiomeric excesses in meteoritic amino acids. *Science*, 275, 951–955.

Damer, B. and Deamer, D. (2020). The hot spring hypothesis for an origin of life. *Astrobiology*, 20, 429–452.

Deamer, D., Damer, B., Kompanichenko, V. (2019). Hydrothermal chemistry and the origin of cellular life. *Astrobiology*, 19, 1523–1537.

Desiraju, G.R. and Steiner, T. (1999). *The Weak Hydrogen Bond in Structural Chemistry and Biology*. Oxford Science Publications Inc., New York.

Di Stefano, R. and Ray, A. (2016). Globular clusters as cradles of life and advanced civilizations. *The Astrophysical Journal*, 827, 54.

van Dishoeck, E.F., Kristensen, L.E., Mottram, J.C., Benz, A.O., Bergin, E.A., Caselli, P., Herpin, F., Hogerheijde, M.R., Johnstone, D., Liseau, R. et al. (2021). Water in star-forming regions: Physics and chemistry from clouds to disks as probed by Herschel spectroscopy. *Astronomy and Astrophysics*, 648, A24.

Ferus, M., Rimmer, P., Cassone, G., Knížek, A., Civiš, S., Šponer, J.E., Ivanek, O., Šponer, J., Saeidfirozeh, H., Kubelík, P. et al. (2020). One-pot hydrogen cyanide-based prebiotic synthesis of canonical nucleobases and glycine initiated by high-velocity impacts on early Earth. *Astrobiology*, 20, 1476–1488.

Forgan, D., Dayal, P., Cockell, C., Libeskind, N. (2017). Evaluating galactic habitability using high-resolution cosmological simulations of galaxy formation. *International Journal of Astrobiology*, 16, 60–73.

Gonzalez, G., Brownlee, D., Ward, P. (2001). The galactic habitable zone: Galactic chemical evolution. *Icarus*, 152, 185–200.

Hirano, S., Hosokawa, T., Yoshida, N., Umeda, H., Omukai, K., Chiaki, G., Yorke, H.W. (2014). One hundred first stars: Protostellar evolution and the final masses. *Astrophys. J.*, 781, 60.

Jiménez-Serra, I., Martín-Pintado, J., Rivilla, V.M., Rodríguez-Almeida, L., Alonso, E.R., Zeng, S., Cocinero, E.J., Martín, S., Requena-Torres, M., Martín-Domenech, R., Testi, L. (2020). Toward the RNA-world in the interstellar medium: Detection of urea and search of 2-amino-oxazole and simple sugars. *Astrobiology*, 20, 1048–1066.

Jørgensen, J.K., Belloche, A., Garrod, R.-T. (2020). Astrochemistry during the formation of stars. *Ann. Rev. Astron. Astrophys.*, 58, 727–778.

Kasting, J.F. and Catling, D. (2003). Evolution of a habitable planet. *Ann. Rev. Astron. Astrophys.*, 41, 429–463.

Kasting, J.F., Whitmire, D.P., Reynolds, R.T. (1993). Habitable zones around main sequence stars. *Icarus*, 101, 108–128.

Kite, E.S. and Ford, E.B. (2018). Habitability of exoplanet water worlds. *The Astrophysical Journal*, 864, 75.

Kopparapu, R.K., Ramirez, R., Kasting, J.F., Eymet, V., Robinson, T.D., Mahadevan, S., Terrien, R.C., Domagal-Goldman, S., Meadows, V., Deshpande, R. (2013). Habitable zones around main-sequence stars: New estimates. *Astrophys. J.*, 765, 131.

Lacki, B.C. (2021). Life in elliptical galaxies: Hot spheroids, fast stars, deadly comets? *The Astrophysical Journal*, 919, 8.

Lazcano, A. (2008). Towards a definition of life: The impossible quest? *Space Sci. Rev.*, 135, 5–10.

Lazcano, A. (2016). Alexandr I. Oparin and the origin of life: A historical reassessment of the heterotrophic theory. *J. Mol. Evol.*, 83, 214–222.

Lineweaver, C.H., Fenner, Y., Gibson, B.K. (2004). The galactic habitable zone and the age distribution of complex life in the Milky Way. *Science*, 303, 59–62.

López-Sepulcre, A., Balucani, N., Ceccarelli, C., Codella, C., Dulieu, F., Theulé, P. (2019). Interstellar formamide (NH_2CHO), a key prebiotic precursor. *ACS Earth and Space Chemistry*, 3, 2122–2137.

Luisi, P.L. (2016). *The Emergence of Life. From Chemical Origins to Synthetic Biology*. Cambridge University Press, Cambridge.

Magill, J. and Galy, J. (2005). *Radioactivity Radionuclides Radiation*. Springer, Berlin/Heidelberg.

Maiolino, R. and Mannucci, F. (2019). De re metallica: The cosmic chemical evolution of galaxies. *Astron. and Astrophys. Rev.*, 27, 1–187.

Mayor, M. and Queloz, D. (1995). A Jupiter-mass companion to a solar-type star. *Nature*, 378, 355–359.

Martins, C.J.A.P. (2017). The status of varying constants: A review of the physics, searches and implications. *Reports on Progress in Physics*, 80, 126902.

McGuire, B.A. (2018). Census of interstellar, circumstellar, extragalactic, protoplanetary disk, and exoplanetary molecules. *Astrophys. J. Suppl. Ser.*, 239, 17.

Miller, S. (1953). A production of amino acids under possible primitive Earth conditions. *Science*, 117, 528–529.

Morrison, I.S. and Gowanlock, M.G. (2015). Extending galactic habitable zone modeling to include the emergence of intelligent life. *Astrobiology*, 15, 683–696.

Natochin, Y.V. (2010). The origin of membranes. *Paleontol. J.*, 44, 860–869.

Nomoto, K., Kobayashi, C., Tominaga, N. (2013). Nucleosynthesis in stars and the chemical enrichment of galaxies. *Ann. Rev. Astron. Astrophys.*, 51, 457–509.

Oparin, A.I. (1924). *Proiskhozhdenje Zhisni*. Moskowski Rabochii, Moscow.

Ozturk, F.S. and Sasselov, D.D. (2022). On the origins of life's homochirality: Inducing enentiomeric excess with spin-polarized electrons. *Proceedings of the National Academy of Sciences*, 119, e2204765119.

Paradise, A., Macdonald, E., Menou, K., Lee, C., Fan, B.L. (2022). ExoPlaSim: Extending the planet simulator for exoplanets. *Monthly Notices of the Royal Astronomical Society*, 511, 3272–3303.

Pascal, R. (2016). Physicochemical requirements inferred for chemical self-organization hardly support an emergence of life in the deep oceans of icy moons. *Astrobiology*, 16, 328–334.

Pearce, B.K.D. and Pudritz, R.E. (2016). Meteorites and the RNA world: A thermodynamic model of nucleobase synthesis within planetesimals. *Astrobiology*, 16, 853–872.

Perryman, M. (2018). *The Handbook of Exoplanets*, 2nd edition. Cambridge University Press, Cambridge.

Pierrehumbert, R.T. (2010). *Principles of Planetary Climate*. Cambridge University Press, Cambridge.

Pizzarello, S. (2016). Molecular asymmetry in prebiotic chemistry: An account from meteorites. *Life*, 6, 18.

Pizzarello, S. and Shock, E. (2017). Carbonaceous chondrite meteorites: The chronicle of a potential evolutionary path between stars and life. *Origins of Life and Evolution of the Biosphere*, 47, 249–260.

Pizzarello, S. and Williams, L.B. (2012). Ammonia in the early solar system: An account from carbonaceous meteorites. *Astrophys. J.*, 749, 161.

Prantzos, N. (2008). On the "galactic habitable zone". *Space Science Rev.*, 135, 313–322.

Quanz, S.P., Ottiger, M., Fontanet, E., Kammerer, J., Menti, F., Dannert, F., Gheorghe, A., Absil, O., Airapetian, V.S., Alei, E. et al. (2022). Large interferometer for exoplanets (LIFE) I. Improved exoplanet detection yield estimates for a large mid-infrared space-interferometer mission. *Astronomy and Astrophysics*, 664, A21.

Ramirez, R.M. and Kaltenegger, L. (2014). The habitable zones of pre-main-sequence stars. *Astrophys. J.*, 797, L25.

Ramirez, R.M. and Kaltenegger, L. (2017). A volcanic hydrogen habitable zone. *Astrophys. J.*, 837, L4.

Ramirez, R.M. and Kaltenegger, L. (2018). A methane extension to the classical habitable zone. *Astrophys. J.*, 858, 72.

Rumble, J.R. (2022). *CRC Handbook of Chemistry and Physics*, 102nd edition. Taylor & Francis Group, London.

Sagan, C. and Salpeter, E.E. (1976). Particles, environments, and possible ecologies in the Jovian atmosphere. *The Astrophysical Journal Supplement Series*, 32, 737–755.

Saladino, R., Di Mauro, E., Garcia-Ruiz, J.M. (2019). A universal geochemical scenario for formamide condensation and prebiotic chemistry. *Chemistry – A European Journal*, 25, 3181–3189.

Schwieterman, E.W., Kiang, N.Y., Parenteau, M.N., Harman, C.E., DasSarma, S., Fisher, T.M., Arney, G.N., Hartnett, H.E., Reinhard, C.T., Olson, S.L. et al. (2018). Exoplanet biosignatures: A review of remotely detectable signs of life. *Astrobiology*, 18, 663–708.

Silva, L., Vladilo, G., Schulte, P.M., Murante, G., Provenzale, A. (2017a). From climate models to planetary habitability: Temperature constraints for complex life. *International Journal of Astrobiology*, 16, 244.

Silva, L., Vladilo, G., Murante, G., Provenzale, A. (2017b). Quantitative estimates of the surface habitability of Kepler-452b. *MNRAS*, 470, 2270–2282.

Simonetti, P., Vladilo, G., Silva, L., Sozzetti, A. (2020). Statistical properties of habitable zones in stellar binary systems. *Astrophys. J.*, 903, 141.

Sojo, V., Herschy, B., Whicher, A., Camprubí, E., Lane, N. (2016). The origin of life in alkaline hydrothermal vents. *Astrobiology*, 16, 181–197.

Spitoni, E., Matteucci, F., Sozzetti, A. (2014). The galactic habitable zone of the Milky Way and M31 from chemical evolution models with gas radial flows. *Monthly Notices of the Royal Astronomical Society*, 440, 2588–2598.

Stojković, N., Vukotić, B., Martinović, N., Ćirković, M.M., Micic, M. (2019). Galactic habitability re-examined: Indications of bimodality. *Monthly Notices of the Royal Astronomical Society*, 490, 408–416.

Toner, J.D. and Catling, D.C. (2019). Alkaline lake settings for concentrated prebiotic cyanide and the origin of life. *Geochimica et Cosmochimica Acta*, 260, 124–132.

Toner, J.D. and Catling, D.C. (2020). A carbonate-rich lake solution to the phosphate problem of the origin of life. *PNAS*, 117, 883–888.

Trifonov, E.N. (2011). Vocabulary of definitions of life suggests a definition. *J. Biomol. Struct. Dyn.*, 29, 259–266.

Varela, F.J., Maturana, H.R., Uribe, R.B. (1974). Autopoiesis: The organization of living systems, its characterization and a model. *Biosystems*, 5, 187–196.

Vladilo, G. (2022). On the role of ^{40}K in the origin of terrestrial life. *Life*, 12, 1620.

Vladilo, G. and Hassanali, A. (2018). Hydrogen bonds and life in the universe. *Life*, 8, 1–22.

Wallace, J.J., Hartman, J.D., Bakos, G.Á. (2020). A search for transiting planets in the globular cluster M4 with K2: Candidates and occurrence limits. *Astron. J.*, 159, 106.

Westall, F. and Brack, A. (2018). The importance of water for life. *Space Science Reviews*, 214, 50.

Westall, F., Hickman-Lewis, K., Hinman, N., Gautret, P., Campbell, K.A., Bréhéret, J.G., Foucher, F., Hubert, A., Sorieul, S., Dass, A.V. et al. (2018). A hydrothermal-sedimentary context for the origin of life. *Astrobiology*, 18, 259–293.

Wolszczan, A. and Frail, D.A. (1992). A planetary system around the millisecond pulsar PSR1257 + 12. *Nature*, 355, 145–147.

Wordsworth, R. and Kreidberg, L. (2022). Atmospheres of rocky exoplanets. *Ann. Rev. Astron. Astrophys.*, 60, 159–201.

Yoshida, S., Grunblatt, S., Price-Whelan, A.M. (2022). Constraining the planet occurrence rate around halo stars of potentially extragalactic origin. *Astron. J.*, 164, 119.

Zahnle, K.J., Lupu, R., Catling, D.C., Wogan, N. (2020). Creation and evolution of impact-generated reduced atmospheres of early Earth. *The Planetary Science Journal*, 1, 11.

Zhu, W. and Dong, S. (2021). Exoplanet statistics and theoretical implications. *Ann. Rev. Astron. Astrophys.*, 59, 291–336.

Ziurys, L.M., Halfen, D.T., Geppert, W., Aikawa, Y. (2016). Following the interstellar history of carbon: From the interiors of stars to the surfaces of planets. *Astrobiology*, 16, 997–1012.

2

Chirality and the Origins of Life

Guillaume LESEIGNEUR and Uwe MEIERHENRICH
Institut de Chimie de Nice, UMR 7272 CNRS, Université Côte d'Azur, France

This chapter is dedicated to the 65th anniversary of Prof. Dr. Wolfgang Weigand, working on prebiotic chemistry at Friedrich Schiller University Jena, Germany.

The notion of chirality is one that is usually not instinctive, yet it is one that is ubiquitous at all scales in the Universe: from the subatomic to the macroscopic. In this chapter, we will introduce the concept of chirality, explain its significance in biology, and give detailed examples of chiral molecules and structures found in life and in the Universe. The homochirality (or molecular asymmetry) of life will be discussed extensively and the current theories aiming to explain it will be described, as its origin remains one of the most fundamental scientific questions today. Its importance in elucidating the origin of life itself, but also in the search for extra-terrestrial life, has made it a useful tool for astrobiology-focused space missions. This is the case for two missions from the European Space Agency (ESA) that we will focus on, one from the past and one from the future. These are Rosetta and ExoMars, two of the growing number of space exploration missions that delve into the molecular origins of life.

2.1. Introduction to chirality

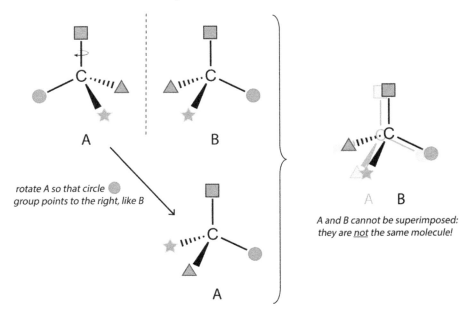

Figure 2.1. *Simplified visualization of the two mirror configurations or enantiomers (A and B) of a chiral molecule. The square, circle, star and triangle represent any four different chemical groups linked to a central carbon. Since the four groups are different, the molecule does not have any internal plane of symmetry and is therefore different from its mirror image. For a color version of this figure, see www.iste.co.uk/ dimauro/firststeps.zip*

NOTE ON FIGURE 2.1.– *B is the mirror image of A. Since they are not identical, A and B are enantiomers. This is because the central carbon is bonded to four different groups (represented by different shapes here), making it asymmetric. Simple lines represent bonds in the plane of the image, wedge-shaped lines represent bonds toward the viewer and dashed lines represent bonds away from the viewer. The carbon is at the center of a tetrahedron, with all its bonded groups at the vertices of this same tetrahedron. This is the most stable way to accommodate four covalent bonds, with the maximum distance between all four electron pairs.*

The term "chiral" comes from the Greek word for "hand" and refers to any object whose mirror image is not the same as itself (meaning objects that do not have mirror symmetry). Chirality is often referred to as handedness, as chiral entities exist in two forms: left and right. Your hands, of course, are chiral: the mirror image of a right hand will be a left hand that is *not* superimposable to the right hand. Any chiral object and its mirror image are therefore fundamentally different, but they possess the same properties. Their interactions with the environment will only differ if this environment, itself, is also chiral. For example, since our hands are chiral entities, we possess a left and a right one, mirror images of one another. Grabbing an achiral object (a sphere, for example) is done in an identical manner with both hands; but grabbing, say, another hand, will result in very different interactions depending on their respective chirality. A handshake between two right hands will have them fit together perfectly, but as most people have experienced before, a left hand shaking a right hand is a different story and simply does not work. It is important to note that an interaction between two left hands will in all points be identical to one between two right hands, only mirrored. Chirality is a property that arises because of our three-dimensional space. It should not be too hard to convince ourselves that a two-dimensional object would not be chiral in 3D space.

Molecules, although extremely small, are three-dimensional entities and can therefore exhibit chirality. In chemistry, the two mirror configurations (left and right) of a chiral molecule are called enantiomers. Chiral molecules possess one or more of what are called chiral centers: these are almost always a tetrahedral (sp^3-hybridized) carbon atom linked to four different groups (Figure 2.1). These are also called asymmetric carbons, as having four different bonds ensures that no internal plane of symmetry is possible. Each of these chiral centers can then be left- or right-handed and are labeled *S* or *R* accordingly (from the Latin "Sinister" and "Rectus", meaning left and right, respectively). For a molecule with several chiral centers, its enantiomer will be the one with the mirror handedness at each asymmetric carbon (each *S* becomes *R* and vice versa).

Two enantiomers of the same molecule have the same properties: they have the same mass, the same size, the same boiling point, the same enthalpy of formation, etc. However, as mentioned above, they will behave differently in a chiral environment or medium, meaning they will interact differently with chiral phenomena. One such phenomenon, which is the one that led to the discovery of molecular chirality by Louis Pasteur, is optical rotation (or circular birefringence). Light is an oscillation of the electromagnetic field: as such this oscillation can be oriented in any direction. Plane-polarized light is defined as having its field oscillations contained in a plane. Pasteur crystallized a salt of a compound known at the time as "racemic acid" (now known as tartaric acid) and noted that it formed two different types of crystals. He observed that plane-polarized light interacted differently with the two crystals after re-dissolving them in water separately. More precisely, they rotated the plane of polarization of light to equal and opposite extents. A balanced mixture of both crystals, however, had no effect on the polarization of the light. This led Pasteur to the conclusion that racemic acid was in fact not a single compound, but a 50:50 mixture of two enantiomers, now called a racemic mixture. A racemic mixture of any molecule will be optically inactive, since the equal and opposite rotations due to each component cancel each other out.

Optical rotation can be explained because plane-polarized light, perhaps surprisingly, is a chiral phenomenon. The plane oscillation can be decomposed as the superposition of two oppositely rotating beams of circularly polarized light: one left and one right.

NOTE.– A rotating vector is not a chiral entity. Circular polarization is a chiral phenomenon because the rotating vectors are moving (at the speed of light) along their spin axis, and are therefore tracing left- and right-handed helices, respectively, along their path.

Each enantiomer will have a preferred interaction with one of the two rotating oscillations, slowing it down through the medium. This equates to mentioning that the refractive indices are different for the left and right circularly polarized beams in a chiral solution, hence the name circular birefringence. The difference in the propagation speed of the two circularly polarized beams results in a rotation of the light's plane of polarization coming out of the medium, hence "optical rotation". The sign of this rotation is one of the ways to assign a handedness to each enantiomer. If the plane of polarized light is rotated clockwise viewed from the point of view of the observer, the enantiomer is dextrorotatory (d) or (+). If the plane of polarized light is rotated counterclockwise (to the left), the molecule is levorotatory (l) or (-). The problem is that the sign of this optical rotation is wavelength dependent, and can also vary with solvent, temperature, concentration and other factors. This makes this notation less than ideal as it is far from an absolute convention.

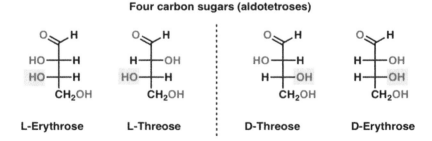

Figure 2.2. *Fischer projection of the enantiomers of the four-carbon sugars: erythrose and threose. For a color version of this figure, see www.iste.co.uk/dimauro/firststeps.zip*

NOTE ON FIGURE 2.2.– *The L sugars have, by convention, the OH group of the bottom chiral carbon on the left. The D sugars have it on the right. For amino acids, we look at the NH2 (amino) group instead. Image taken from masterorganicchemistry.com with permission from James Ashenhurst.*

One of the oldest conventions to name enantiomers is the L and D notation (small uppercase, not to be confused with *l* and *d*), which refers specifically to their absolute configuration, that is the precise arrangement of their atoms in space. It is based on the Fischer projection of a molecule (Figure 2.2) and has the advantage of being immutable. It is still used today for amino acids (the monomers of proteins) and for sugars (as constituents of nucleotides, the monomers of the backbone of DNA and RNA). The L/D notation has the advantage of remaining a single letter even when there are several chiral centers in the molecule in question (which is the case for sugars). This is in contrast with the *R/S* notation that requires specifying the absolute configuration of all asymmetric carbons.

2.2. The asymmetry of life

Life is fundamentally asymmetric. Homochirality (the use of only one of the possible chiral configurations) is already observed in nature at the molecular level, with proteins being polymers of exclusively L-amino acids (Meierhenrich 2008). This homochirality in the monomeric building blocks of proteins leads to homochirality in higher order structures such as the right-handed α-helix (secondary structure) and the fold (tertiary structure) that is unique to each different protein in its native state (Barron 2008). Similarly, DNA and RNA are made of a backbone of exclusively D-sugars, respectively, deoxyribose and ribose. This induces a left-handedness to the nucleotides, that in the case of DNA creates the right-handed double helix of its strands. This continues with the supercoiling of DNA exhibiting left-handed homochirality. What we observe is a cascading effect of the monomeric homochirality of biomolecules that creates an alternating handedness of the molecular complexes as we move through different scales of the biological machinery (Tverdislov et al. 2017). Therefore, in nature, homochirality occurs at all scales (Hegstrom and Kondepudi 1990). In fact, any chiral structure incorporated in a biological system will display homochirality. This is not limited to amino acids, sugars and their resulting macro-structures. Homochirality is essential for an efficient biochemistry, much like the universal adoption of right-handed screws in engineering. It does not matter which hand is used as long as only one hand is used: there is no "better" handedness, the benefit arises only when one is universally adopted. Another example is one that we mentioned before: the right-hand handshake. Again, the interaction between two left hands is in all points identical to the one between two right hands, only mirrored. That only one of these is the norm culturally is a matter of increasing the efficiency of everyday interactions. This begs the question: why this handedness and not the other? While our screw and handshake conventions can be attributed to the fact that 85% of the population is right-handed, this simply pushes the question back. Why are the vast majority of humans right-handed? Indeed, there has never been any report of a human population in which left-handed individuals predominate (Uomini 2009).

To complicate the matter, a chiral entity does not always interact preferentially with another one of the same handedness. A right hand prefers to shake another right hand, but prefers to hold a left hand when standing side by side. The interaction of molecules of the same nature (protein–protein, etc.) mainly occurs in the case of the same sense of chirality, either L–L or D–D, and for molecules of different types (e.g. DNA–protein), in the case of different senses of chirality, either D–L or L–D. These chirally discriminating interactions produce the connections between the L-amino acids and D-sugars in life. Based on the examples above, we understand that a "mirror" biology, made from D-amino acids and L-sugars, would be just as viable. This is something that was recognized very early (Fischer 1894), by none other than the eponym of the Fischer projection (Figure 2.2). With his "lock and key" hypothesis, Fischer understood that a chiral molecule could discriminate between two mirror forms of another chiral molecule, as although they have the same connectivity (each atom of the molecule is linked to the same atoms), they have different spatial arrangements of their bonds. In other words, the mirror image of the key to a chiral lock would not be able to open it. The sheer size of the molecules used in (our carbon-based) biology means that chiral centers (asymmetric carbons) are ubiquitous. Thus, the overwhelming majority of biomolecules are chiral and will invariably be found in homochiral form.

Importantly, homochirality is only maintained inside of a biological system. Outside of one, the second law of thermodynamics dictates a perpetual tendency to racemize (go back to a 50:50 mix of both enantiomers). Life, in a way, is a constant fight against racemization, against unceasingly increasing entropy. The importance of homochirality in biological systems is underlined by the fact that the "wrong" hand often has destructive effects: the preservation of homochirality is a key component of protein function that is essential to maintain homeostasis across the cell, tissue and organ level (Banreti et al. 2022). In fact, the body contains special enzymes called D-amino acid oxidases to eliminate amino acids of the wrong hand (D'Aniello et al. 1993). Almost all metabolic processes in the cell need enzyme catalysis in order to occur at rates fast enough to sustain life. Enzymes are protein molecules that fold up into complex three-dimensional structures, providing an active site which reactants can fit into. Since enzymes' specificity comes from their unique three-dimensional structure, only molecules of the right shape can participate in these catalyzed reactions. In this way, enzymes, built out of L-amino acids, can synthesize more L-amino acids. In general, however, chemical processes only produce homochiral products starting from other homochiral molecules (Bailey 2000). It follows that homochirality is not a consequence but rather a pre-condition for life. This raises the question of how and when homochirality originated.

2.3. The origin of homochirality

The implausibility of life without chirality raises fundamental questions. Perhaps the most elementary one is why were L-amino acids and D-sugars chosen, and not the opposite? This is a question that remains, to this day, unsolved. Many hypotheses have been formulated, either deterministic or stochastic. A deterministic theory entails that there are nonrandom external asymmetric forces that would yield products of a pre-determined chirality. In that sense, the incorporation of L-amino acids and D-sugars in life would be universal, or at least the only way it could have happened on Earth. Stochastic (or non-deterministic) theories propose that the choice of L-amino acids and D-sugars was due to chance mechanisms, without the need for an inherent nudge toward one or the other handedness.

We then require a fast enough mechanism for the formation of the homochiral biopolymers, leading us from our initial enantiomeric excess (due to chance or not) to homochirality. A significant peril to prebiotic chirality, of course, was racemization (Bonner 1993). The racemization of amino acids on the modern Earth is a rapid geochemical reaction; it converts biological L-amino acids into racemic mixtures in 103 years at 50°C and 106 years at 0°C. Therefore, any homochiral amino acids on the primitive Earth must have been incorporated into important biomolecules and somehow protected from racemization on these timescales (Bada 1995).

2.3.1. Stochastic theories

One of the first "chance" theories was the proposition of a theoretical spontaneous symmetry breaking mechanism, where one enantiomer is a catalyst for its own production, while being an anti-catalyst for its antipode (Frank 1953). From this system, any random fluctuation causing a deviation from a 50:50 or racemic mixture would result in the dominance of one of the enantiomers through this positive feedback loop. In this scenario, the racemic state is therefore an unstable equilibrium point. This was shown experimentally in a non-aqueous environment, in a chemical system with a starting enantiomeric excess of 2%, meaning a 51:49

mixture (Soai et al. 1995). However, while 2% may seem small, it is significant and bigger than even what we can expect from most proposed deterministic forces (see section 2.3.2). Purely stochastic fluctuations can be quantified with simple statistics: for N molecules in a racemate, the most likely state is one where the dominant enantiomer will have $(N/2) + \sqrt{N}$ molecules. For this to represent 2% of N, we would need our system to be no bigger than 10,000 molecules.

Chiral symmetry breaking has been shown to occur during crystallization of sodium chlorate from an aqueous solution while the solution is stirred (Kondepudi et al. 1990). This represents a total spontaneous resolution on a macroscopic level, brought about by autocatalysis and competition between L- and D-crystals. An asymmetric adsorption of enantiomers can then occur on homochiral crystals such as quartz (Karagunis and Coumoulos 1938; Bonner et al. 1974), or other mineral surfaces such as calcite (Hazen et al. 2001). Some have proposed the origin of molecular chiral asymmetry is associated with fractionation of enantiomers at the ocean-atmosphere non-equilibrium boundary during the origination of the predecessors of living cells (Tverdislov et al. 2007). It has also been proposed that homochirality arose not at the monomer level, but during the polymerization step (Zepik et al. 2007; Chen and Ma 2020). Most theories that aim to explain how chirality can emerge from a non-chiral environment imply far-from equilibrium conditions, with continuous flows of energy. Energy flows are required as the driving force for self-organizing processes (i.e. crystallizations, polymerizations and supramolecular organizations) where chiral dissipative structures are formed (Weissbuch et al. 2005).

The main argument of chance mechanism theories is that the origin of life must be a rare event on our timescale, and given the massive generality of statistical arguments, the spontaneous generation of homochirality is its most likely origin, more likely than the chance of life arising at all. In that sense, there is no *need* to recur to deterministic forces or astrophysical events (Siegel 1998). It should be noted that the stochastic theories discussed above are not in contradiction with the possibility of a deterministic chiral bias at a larger level. Indeed, the nature of stochastic theories requires them to also be a chiral amplification mechanism for the random fluctuations that allegedly led to homochirality. This amplification mechanism from small enantiomeric excess to homochirality is also necessary for deterministic theories, which we will now present.

2.3.2. Deterministic theories

Deterministic theories are looking for inherent asymmetries at the larger scales of the Universe, or even within its fundamental laws. Indeed, out of the four fundamental forces (electromagnetic, weak, strong and gravity), one was shown to be chiral: the weak force (Lee and Yang 1956). One of the consequences of this chirality was the asymmetric β-decay of radionuclides, observed soon after (Wu et al. 1957). The details are out of the scope of this chapter, but this established that fermions (half-odd-integer spin particles, which include the electron) exist in two states of opposite helicity. While they participate equally in the electromagnetic, strong and gravitational interactions, they do not participate equally in the weak interaction, which is said to violate parity. Indeed, left-handed electrons participate in the weak interaction preferentially compared with right-handed electrons, while right-handed positrons (the anti-particle of the electron) are preferred over left-handed positrons. In that sense, the true enantiomer of an L-amino acid is a D-amino acid made of anti-matter (MacDermott 1995). This very profound result shows that the Universe is intrinsically "left-handed" because of an all-pervading asymmetry: it is made of matter and not anti-matter. Therefore, the main argument of parity violation theories is that this universal chiral influence could determine biomolecular chirality through what is called the parity-violating energy difference (PVED) between left- and right-handed molecules. This arises from weak neutral current interactions, mediated by the $Z°$ boson, between electrons and neutrons. However, these PVEDs are extremely small, orders of magnitude smaller than statistical fluctuations. The biggest result for this theory has been that the PVED between two enantiomers was shown to almost systematically favor the populations of "natural" enantiomers (L-amino acid, D-sugars and their polymers) over their "unnatural" mirror image to the extent of about 1 part in 10^{17} (Mason 1984; Mason and Tranter 1985). This means that the biological versions of chiral monomers and polymers are ever so slightly more stable than their non-biological counterparts due to the asymmetry of the weak interactions. Theoretical mechanisms of amplification for this minute but systematic chiral force have been proposed (Kondepudi and Nelson 1985; Kondepudi 1987), but the almost imperceptible magnitude of the effects at play and the lack of experimental evidence to date for any efficacy of PVEDs in producing enantiomeric excesses ensure that this theory remains in the realm of supposition.

While the electromagnetic force is not chiral in itself, the polarization state of photons is chiral. We have seen that circularly polarized light (CPL) is an electromagnetic wave whose electric vector traces a left- (l-CPL) or right-handed spiral (r-CPL) along its direction of propagation (see section 2.1). This helicity and therefore chirality of CPL makes it distinguish between left- and right-handed molecules in its interactions. We have explained the physics behind circular birefringence or optical rotation, but another phenomenon falling under the umbrella of the more general term "optical activity" is circular dichroism. While circular birefringence is the difference in *refraction* (and the associated speed of light), circular dichroism is the difference in *absorption* of left and right CPL by the respective enantiomers. Since the rate of a photochemical reaction is determined by the amount of light absorbed, this means that under CPL, reaction rates will be different for left- and right-handed enantiomers. Importantly this means that asymmetric photolysis or synthesis, the selective destruction or formation of an enantiomer by CPL, would be possible in space (Griesbeck and Meierhenrich 2002; Meierhenrich and Thiemann 2004; Meinert et al. 2011). Indeed, high circular polarization in the infrared range (a maximum of 17%) has been observed in the massive star forming region of the Orion nebula (Bailey et al. 1998; Fukue et al. 2010). The origin of this polarization at such scales has yet to be confirmed, but the most convincing theory so far is the induction of CPL by scattering processes from micron-size interstellar dust grains (Fukushima et al. 2020). However, photolysis of amino acids requires UV radiation, rather than the infrared radiation that has so far been observed. Unfortunately, UV radiation cannot be directly observed as it is blocked by the same dust grains present along our line-of-sight toward these regions of high circular polarization. Still, theoretical investigations of mechanisms by which ultraviolet circular polarization could be produced in star formation regions have shown that dichroic extinction of linearly polarized light by non-spherical dust grains could produce a net circular polarization of about 10% (Lucas et al. 2005). The non-sphericity of the dust grains creates an asymmetry as they align with the local magnetic field (Hough and Aitken 2003), asymmetry that is then inherited by the absorbed and scattered photons.

Circular dichroism spectra of gas phase amino acids in the UV range were obtained and showed that CPL irradiation can induce an enantiomeric excess of about 2% for a photolysis fraction of 99.99% (Meinert et al. 2022). We see that we require the destruction of the overwhelming majority of a sample, even under irradiation from pure *l*- or *r*-CPL, to have a measurable effect. Still, the CPL theory has gained a lot of traction after significant enantiomeric excesses of amino acids and sugars were found in extraterrestrial samples (Garcia et al. 2019). From several samples of the Murchison meteorite and one from the Murray meteorite, L-enantiomeric excesses of the amino acid isovaline were found to range from 0 to 15.2%. Individual D- and L-isovaline $\delta^{13}C$ values were found to be about +18‰, with no significant differences between the two enantiomers to suggest terrestrial contamination (Pizzarello et al. 2003). This is the largest enantiomeric excess measured to date for a biologically rare meteoritic amino acid, and its magnitude raises doubts that CPL irradiation could be the sole cause of amino acids chiral asymmetry in meteorites. Enantiomeric enrichment processes possible on the meteorites' parent bodies have been proposed (Glavin and Dworkin 2009). Other amino acids were found with enantiomeric excesses of the L form, including non-proteinogenic amino acids, which virtually removed a contamination from the biosphere as a possibility for this asymmetry (Cronin and Pizzarello 1997; Cronin and Pizzarello 1999). Isotopic measurements showed that individual amino acid enantiomers from Murchison are enriched in ^{15}N relative to their terrestrial counterparts, confirming an extraterrestrial source for an L-enantiomer excess in the solar system that may predate the origin of life on Earth (Engel and Macko 1997). More recently, it has been found that sugar monoacids (aldonic acids) exhibited important enantiomeric excesses (up to 82%) of the "natural" D-enantiomer in several carbonaceous meteorites. The fact that this was observed in both common and rare sugars indicates again an extraterrestrial origin. This and concurring $^{13}C/^{12}C$ isotopic measurements hint at a source from abiotic chemistry, and significantly lessen the possibility of contamination (Cooper and Rios 2016). We see that for both amino acids and sugars, enantiomeric excesses found in meteorites are in accordance with the handedness of life on Earth. This could then be an explanation for why these configurations were the ones chosen as the basis for – and amplified to homochirality by – the first self-replicating macromolecules.

In summary, mirror symmetry breaking events seem very likely in self-organizing chemical systems. It can then be rationalized that homochirality was most probably a general feature of the first biopolymers and evolved together with their self-replication capabilities and the lipid compartments in which they were encapsulated (Ruiz-Mirazo et al. 2014). There are many possible phenomena that have been proposed for the origin of homochirality, making it more and more difficult to believe in a single overarching event for its emergence. It is likely that fundamentally asymmetrical forces played a role in the appearance of a chiral bias and were then amplified by one or most of the many proposed stochastic mechanisms. The biggest remaining question is whether the deterministic factors discussed above are significant enough to dictate the chirality of life universally, or if the chance to end up in our "L-amino acids and D-sugars system" was close enough to 50% that they did not matter. It seems that without the detection of a mirror life form (one that uses D-amino acids and/or L-sugars) in our solar system, there is no possible argument to rule out an astrophysical and deterministic origin for homochirality. This has become one of the main incentives for embarking chiral analytical instruments on astrobiology-focused space missions (Glavin et al. 2020).

2.4. Space missions and the search for life and its origins

Looking for extraterrestrial chiral molecules can help answer two fundamental questions: first, by looking at asteroids and comets, we can observe if any chiral bias is introduced already in the interplanetary medium through abiogenic processes, which would hint at a possible origin of the asymmetry of chiral monomers. Indeed, through impact events (meteorites), they would have brought organic molecules of prebiotic interest with an already embedded chiral asymmetry (Meierhenrich 2014). Therefore, meteorites currently falling on Earth are an extremely important tool to understand the chemical composition of the early solar system and consequently of the early Earth, but they are prone to thermal degradation from the impact and to contamination from the biosphere between their fall and their retrieval.

A much more expensive possibility is to send rovers directly to these asteroids and comets to analyze their composition in situ, thereby obtaining the most unaltered samples conceivable. Indeed, asteroids and even more so comets are believed to carry the most pristine material preserved in the solar system, largely undisturbed since their formation more than 4.5 billion years ago. According to the Greenberg model of comet formation, small molecules condense from the gas phase on the cold surface of interstellar dust grains, thereby forming an icy mantle: ices rich in water (~80%) and organic molecules (Greenberg 1984). Comets are then an almost pure aggregate of these icy grains, hence the term "dirty snowball" often used to qualify them. This relatively simple model of gas condensation on dust grains can be reproduced in laboratory experiments, in what have been called simulated interstellar ices (Myrgorodska et al. 2015). When irradiated with UV light simulating our Sun, amino acids and sugars have been detected in these interstellar ice analogues (Muñoz Caro et al. 2002; Meinert et al. 2016). More interestingly, when the small precursor molecules were irradiated with CPL during condensation, an enantiomeric excess was found in the synthetized amino acids (de Marcellus et al. 2011; Modica et al. 2014). This proved the feasibility and detectability of the asymmetric photochemical synthesis of amino acids from achiral interstellar precursors under pure *l*- or *r*-CPL irradiation. Confirmation of these models through space exploration missions remains the biggest challenge.

The most recent endeavor in cometary exploration, and the only one that included an instrument capable of chiral separation, was the Philae rover, which was the lander from the European Space Agency's (ESA) Rosetta space mission that visited comet 67P/Churyumov-Gerasimenko (see section 2.4.1). Even more recent and with analyses still ongoing, is the sample-return mission Hayabusa 2 from JAXA, the Japanese space agency. The mission went to asteroid Ryugu to collect a sample from its surface and returned in December 2020 with 5.4 g of material, more than 50 times what was expected. Finally, the most recent endeavor in asteroid sample-return is the OSIRIS-Rex mission from the North American Space Agency (NASA), which is currently on its way back with what is believed to be upward of 400 g of material from the asteroid Bennu and is to arrive on Earth in September 2023. This will allow scientists all over the world to use the latest analytical techniques to characterize the composition of asteroids with greater precision than ever.

The second fundamental question that extraterrestrial chiral molecules can help answer is on the presence of life elsewhere in the Universe. Since homochirality is only maintained inside of a biological system, it is what is called a biosignature. Finding enantiomeric excesses of any molecules on other planets or moons would therefore be a possible indication of a past or present biological activity. Here again, the only options are either sample-return space missions or space missions capable of chiral separation in an embarked analytical laboratory. The next major space exploration mission capable of chiral discrimination is the latter and will be ESA's ExoMars mission, with its Rosalind Franklin rover set to land on the red planet in 2028 (see section 2.4.2). Both Philae from Rosetta and Rosalind Franklin from ExoMars will be equipped with an instrument capable of chiral separation in the form of a gas chromatographic column (the CP-Chirasil-Dex CB in both cases) onboard a miniaturized GC-MS (gas chromatograph–mass spectrometer) analytical laboratory.

2.4.1. *Rosetta*

Rosetta arrived at comet 67P/Churyumov-Gerasimenko in August 2014. The spacecraft orbited the nucleus of the comet (Figure 2.2) at a distance of only a few kilometers and was therefore well within its coma (the gaseous envelop of a comet). For chemical characterization, the orbiter was equipped with mass spectrometers that identified many organic molecules within the cometary coma, including glycine, the simplest and only achiral amino acid (Altwegg et al. 2016). In any case, a mass spectrometer alone cannot differentiate between enantiomers. Attached to the orbiter was Philae, the rover that then detached to go land on the nucleus of a comet for the first time in history. Onboard the Philae lander was the COSAC instrument (Goesmann et al. 2007) that with a chiral chromatographic column before the mass spectrometer detector had the possibility of separating and identifying the enantiomers of a large variety of molecules, including amino acids (Thiemann and Meierhenrich 2001). Unfortunately, Philae's landing did not go as planned due to a malfunction of the harpoons that were supposed to anchor the probe to the surface (Ulamec et al. 2016). The lander bounced several times before coming to a rest in an awkward position on the nucleus. A single mass spectrum was taken mid-bounce and allowed the tentative detection of 16 molecules (Goesmann et al. 2015), updated to 12 main contributions with more recent deconvolution algorithms (Leseigneur et al. 2022c). These represent molecules that entered the mass spectrometer chamber passively through the exhaust pipe of the instrument during the bounce, and therefore represent the first ever in situ analysis of a cometary nucleus. Unfortunately, due to this unexpected sampling method, only molecules too small to exhibit chirality were found.

The chiral experiment of COSAC was supposed to happen with the lander safely positioned on the surface, with a foraging drill to feed crushed samples from the surface. The final position of Philae meant that it was unable to properly recharge its batteries, so a first and last science sequence was started, and a chiral column was chosen as it represented the biggest potential scientific return. The COSAC instrument performed the science sequence nominally, but could not unequivocally identify any molecule from this gas chromatograph due to a very poor signal-over-noise ratio. Indeed, the drill is believed to not have reached the surface due to the position of Philae with respect to the local ground (Di Lizia et al. 2016). What little material may have entered the injection port of COSAC would be ambient gas that stuck to the drill as it deployed. The detection of the achiral molecule ethylene glycol $(CH_2OH)_2$ at trace level within this dataset was proposed recently (Leseigneur et al. 2022b), which would confirm the ubiquity of this molecule in cometary environments (Crovisier et al. 2004; Biver et al. 2014; Biver et al. 2021) and in the interstellar medium (Brouillet et al. 2015; Li et al. 2017). In any case, no chiral molecules could be identified in COSAC's only gas chromatogram; the first extraterrestrial enantiomeric separation will therefore have to wait.

Figure 2.3. *Single frame Rosetta navigation camera image of 67P/Churyumov-Gerasimenko taken on July 7, 2015 from a distance of 154 km from the comet center. The image has a resolution of 13.1 m/pixel and measures 13.4 km across. Image: ESA – European Space Agency*

2.4.2. *ExoMars*

The Rosalind Franklin rover (Figure 2.4) of ESA's ExoMars mission will contain the MOMA (Mars Organic Molecule Analyzer) instrument (Goesmann et al. 2017), a GC-MS that has a strong heritage from COSAC of Rosetta. It will carry the same chiral chromatographic column capable of separating enantiomers of a large range of molecules, including amino acids, of course (Guzman et al. 2020). Mars is a planet that shared a similar early geological history with Earth, particularly during the time when life is supposed to have appeared on our planet. Indeed, it is likely that 4.45 billion years ago, early Mars also developed a global ocean (or large bodies of water) enveloped in a very dense and mostly CO_2 atmosphere (Elkins-Tanton 2011). Although, due to its lower gravity, they eventually evaporated, liquid water is expected to have lasted several hundred million years on the Martian surface (Di Achille and Hynek 2010). This means conditions compatible with life should have remained until around 4 billion years ago, after which the habitability of the red planet lessened significantly to finally become what it is today: a very cold, desert-like planet. Therefore, even if life emerged then, it is likely that it still went extinct a very long time ago.

Looking for traces of past biologic activity is the main mission goal of ExoMars, and to do so it will be the first Martian rover capable of drilling down to a depth of 2 m, while previous rovers were only able to scrape the surface. This is of immense scientific value because of the very thin atmosphere there; the surface is bombarded with UV radiation from the Sun and is therefore completely sterile. Having a subsurface drill greatly increases the probability of collecting well-preserved material for analysis. The absence of plate tectonics on Mars increases the chances that rapidly buried ancient sedimentary rocks (possibly hosting microorganisms), may have been spared from thermal alteration and shielded from ionizing radiation damage. The obvious targets for molecular biosignatures are of course the ensemble of primary biomolecules such as amino acids, nucleic acids, carbohydrates and so on. Discriminating between the enantiomers of these molecules and detecting a homochiral presence of either of them would be almost irrefutable proof of extant life on the red planet, but unfortunately they degrade quickly once microbes die. They are therefore the best tools for detecting current biologic activity, but due to relatively fast racemization and overall degradation they are not expected to survive on a geological timescale. Lipids, however, are also biologically essential components (most importantly of cell membranes), but are stable for billions of years when buried (Brocks et al. 1999; Georgiou and Deamer 2014). It is the recalcitrant hydrocarbon backbone that is responsible for the high-preservation potential of lipid-derived biomarkers relative to that of other biomolecules

(Eigenbrode 2008). This hydrocarbon backbone is chiral for archaeal cells, and therefore it has a homochiral presence in biology. The same argumentation is true for the side chain of the main photosynthetic pigment chlorophyll that has the same relative configuration as the archaeal membrane (Figure 2.5). It has recently been pointed out that having access to the chirality of hydrocarbons, although far from trivial, should be doable with the MOMA instrument (Leseigneur et al. 2022a). It has been explicitly stated by the ExoMars science team that, for any family of molecules, a measured enantiomeric excess going toward homochirality would be the biggest indicator of a biologic activity (Vago et al. 2017). Given the history of Mars, being able to do so on chiral hydrocarbons would be what enables us to probe for long extinct organisms; to look back in time the furthest, all the way to the planet's infancy, and possibly the beginnings of life there.

Outside of Mars, the most promising recent findings for astrobiology-focused space exploration missions are ocean worlds. We are finding more and more moons of Saturn and Jupiter with global subsurface water oceans. These are found several kilometers deep, but are of course an environment of great interest for the potential evolution of living organisms. One of the earliest upcoming space missions to an ocean world is DragonFly that will land a rotorcraft on Titan, Saturn's largest moon, in the mid-2030s (Barnes et al. 2016). It will be equipped with a chiral GC-MS, with heritage from MOMA and COSAC, to look for chemical biosignatures indicative of past or extant biological processes at its surface, as we expect mixing between the subsurface ocean and the surface of Titan through cryovolcanic activity. An Orbilander concept has been proposed to explore another moon of Saturn with a subsurface ocean: Enceladus. Indeed, its surface vents ice into space through water plumes, believed to be similar in composition to comets. The mission project has not yet been accepted, but is in NASA's plans for one of the next major space explorations. The instrument suite has not yet been fully defined, but it is now clear that any mission that has as a scientific goal to look for traces of extraterrestrial life must include instruments capable of chiral separation. Chiral gas chromatography has proved its viability in space, but other smaller instruments such as an embarked polarizer have been proposed, that could be used alone for smaller missions, as independent corroboration of a GC-MS, or even as the detector of a gas chromatograph (MacDermott et al. 1996). After the successful Rosetta mission, a very early draft of the AMBITION project for a sample-return mission to a comet nucleus has been put forward (Bockelée-Morvan et al. 2021). These scientific endeavors are all part of an aspiration to understand the origin of homochirality of life on Earth and perhaps elsewhere in our solar system. They are constantly expanding the boundaries of knowledge on our own history and our Universe.

Chirality and the Origins of Life 47

Figure 2.4. *Artistic view of the Rosalind Franklin rover on Mars, with the foldable drill in the black cylinder in the front of the rover. The rover weighs 310 kg and is expected to launch in 2028 for a landing on Mars between 2029 and 2030, depending on the final launch details. Image: ESA/ATG medialab. For a color version of this figure, see www.iste.co.uk/dimauro/firststeps.zip*

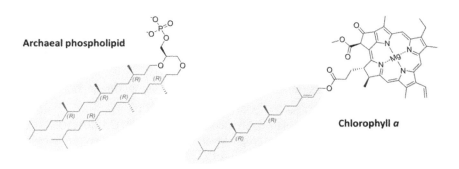

Figure 2.5. *Visualization of an archaeal phospholipid (left) and chlorophyll a (right)*

NOTE ON FIGURE 2.5.– *Highlighted are the homologous chiral hydrocarbon side chains for both. Note that all asymmetric carbons are of the same handedness (R from the R/S notation), hinting at a common enzymatic origin for both. In chlorophyll, this side chain is ester-linked to the active part of the molecule, which is the porphyrin ring that here serves as a photoreceptor. In the archaeal membrane lipid, the side chains are ether-linked to the hydrophilic glycerol-phosphate head.*

2.5. References

Altwegg, K., Balsiger, H., Bar-Nun, A., Berthelier, J.-J., Bieler, A., Bochsler, P., Briois, C., Calmonte, U., Combi, M.R., Cottin, H. et al. (2016). Prebiotic chemicals – amino acid and phosphorus – in the coma of comet 67P/Churyumov-Gerasimenko. *Science Advances*, 2(5), e1600285.

Bada, J.L. (1995). Origins of homochirality. *Nature*, 374(6523), 594–595.

Bailey, J. (2000). Chirality and the origin of life. *Acta Astronautica*, 46(10), 627–631.

Bailey, J., Chrysostomou, A., Hough, J.H., Gledhill, T.M., McCall, A., Clark, S., Ménard, F., Tamura, M. (1998). Circular polarization in star-formation regions: Implications for biomolecular homochirality. *Science*, 281(5377), 672–674.

Banreti, A., Bhattacharya, S., Wien, F., Matsuo, K., Réfrégiers, M., Meinert, C., Meierhenrich, U., Hudry, B., Thompson, D., Noselli, S. (2022). Biological effects of the loss of homochirality in a multicellular organism. *Nature Communications*, 13(1), 7059.

Barnes, J.J., Kring, D.A., Tartèse, R., Franchi, I.A., Anand, M., Russell, S.S. (2016). An asteroidal origin for water in the Moon. *Nature Communications*, 7(1), 11684.

Barron, L.D. (2008). Chirality and life. *Space Science Reviews*, 135(1–4), 187–201.

Biver, N., Bockelée-Morvan, D., Debout, V., Crovisier, J., Boissier, J., Lis, D.C., Russo, N.D., Moreno, R., Colom, P., Paubert, G. et al. (2014). Complex organic molecules in comets C/2012 F6 (Lemmon) and C/2013 R1 (Lovejoy): Detection of ethylene glycol and formamide. *Astronomy & Astrophysics*, 566, L5.

Biver, N., Bockelée-Morvan, D., Boissier, J., Moreno, R., Crovisier, J., Lis, D.C., Colom, P., Cordiner, M.A., Milam, S.N., Roth, N.X. et al. (2021). Molecular composition of comet 46P/Wirtanen from millimetre-wave spectroscopy. *Astronomy & Astrophysics*, 648, A49.

Bockelée-Morvan, D., Filacchione, G., Altwegg, K., Bianchi, E., Bizzarro, M., Blum, J., Bonal, L., Capaccioni, F., Choukroun, M., Codella, C. et al. (2021). AMBITION – Comet nucleus cryogenic sample return. *Experimental Astronomy* [Online]. Available at: https://link.springer.com/10.1007/s10686-021-09770-4 [Accessed 22 June 2022].

Bonner, W.A. (1995). Chirality and life. *Origins of Life and Evolution of the Biosphere*, 25(1), 175–190.

Bonner, W.A., Kavasmaneck, P.R., Martin, F.S., Flores, J.J. (1974). Asymmetric adsorption of alanine by quartz. *Science*, 186(4159), 143–144.

Brocks, J.J., Logan, G.A., Buick, R., Summons, R.E. (1999). Archean molecular fossils and the early rise of eukaryotes. *Science*, 285(5430), 1033–1036.

Brouillet, N., Despois, D., Lu, X.-H., Baudry, A., Cernicharo, J., Bockelée-Morvan, D., Crovisier, J., Biver, N. (2015). Antifreeze in the hot core of Orion-First detection of ethylene glycol in Orion-KL. *Astronomy & Astrophysics*, 576, A129.

Chen, Y. and Ma, W. (2020). The origin of biological homochirality along with the origin of life. *PLOS Computational Biology*, 16(1), e1007592.

Cooper, G. and Rios, A.C. (2016). Enantiomer excesses of rare and common sugar derivatives in carbonaceous meteorites. *Proceedings of the National Academy of Sciences*, 113(24), E3322–E3331.

Cronin, J.R. and Pizzarello, S. (1997). Enantiomeric excesses in meteoritic amino acids. *Science*, 275(5302), 951–955.

Cronin, J.R. and Pizzarello, S. (1999). Amino acid enantiomer excesses in meteorites: Origin and significance. *Advances in Space Research*, 23(2), 293–299.

Crovisier, J., Bockelée-Morvan, D., Biver, N., Colom, P., Despois, D., Lis, D.C. (2004). Ethylene glycol in comet C/1995 O1 (Hale-Bopp). *Astronomy & Astrophysics*, 418(3), L35–L38.

D'Aniello, A., D'Onofrio, G., Pischetola, M., D'Aniello, G., Vetere, A., Petrucelli, L., Fisher, G.H. (1993). Biological role of D-amino acid oxidase and D-aspartate oxidase. Effects of D-amino acids. *Journal of Biological Chemistry*, 268(36), 26941–26949.

Di Achille, G. and Hynek, B.M. (2010). Ancient ocean on Mars supported by global distribution of deltas and valleys. *Nature Geoscience*, 3(7), 459–463.

Di Lizia, P., Bernelli-Zazzera, F., Ercoli-Finzi, A., Mottola, S., Fantinati, C., Remetean, E., Dolives, B. (2016). Planning and implementation of the on-comet operations of the instrument SD2 onboard the lander Philae of Rosetta mission. *Acta Astronautica*, 125, 183–195.

Eigenbrode, J.L. (2008). Fossil lipids for life-detection: A case study from the early earth record. *Space Science Reviews*, 135(1), 161–185.

Elkins-Tanton, L.T. (2011). Formation of early water oceans on rocky planets. *Astrophysics and Space Science*, 332(2), 359–364.

Engel, M.H. and Macko, S.A. (1997). Isotopic evidence for extraterrestrial non-racemic amino acids in the Murchison meteorite. *Nature*, 389(6648), 265–268.

Fischer, E. (1894). Einfluss der configuration auf die wirkung der enzyme. *Berichte der deutschen chemischen Gesellschaft*, 27(3), 2985–2993.

Frank, F.C. (1953). On spontaneous asymmetric synthesis. *Biochimica et Biophysica Acta*, 11, 459–463.

Fukue, T., Tamura, M., Kandori, R., Kusakabe, N., Hough, J.H., Bailey, J., Whittet, D.C.B., Lucas, P.W., Nakajima, Y., Hashimoto, J. (2010). Extended high circular polarization in the orion massive star forming region: Implications for the origin of homochirality in the solar system. *Origins of Life and Evolution of Biospheres*, 40(3), 335–346.

Fukushima, H., Yajima, H., Umemura, M. (2020). High circular polarization of near-infrared light induced by micron-sized dust grains. *Monthly Notices of the Royal Astronomical Society*, 496(3), 2762–2767.

Garcia, A.D., Meinert, C., Sugahara, H., Jones, N.C., Hoffmann, S.V., Meierhenrich, U.J. (2019). The astrophysical formation of asymmetric molecules and the emergence of a chiral bias. *Life*, 9(1), 29.

Georgiou, C.D. and Deamer, D.W. (2014). Lipids as universal biomarkers of extraterrestrial life. *Astrobiology*, 14(6), 541–549.

Glavin, D.P. and Dworkin, J.P. (2009). Enrichment of the amino acid L-isovaline by aqueous alteration on CI and CM meteorite parent bodies. *Proceedings of the National Academy of Sciences*, 106(14), 5487–5492.

Glavin, D.P., Burton, A.S., Elsila, J.E., Aponte, J.C., Dworkin, J.P. (2020). The search for chiral asymmetry as a potential biosignature in our solar system. *Chemical Reviews*, 120(11), 4660–4689.

Goesmann, F., Rosenbauer, H., Roll, R., Szopa, C., Raulin, F., Sternberg, R., Israel, G., Meierhenrich, U., Thiemann, W., Munoz-Caro, G. (2007). Cosac, the cometary sampling and composition experiment on Philae. *Space Science Reviews*, 128(1–4), 257–280.

Goesmann, F., Rosenbauer, H., Bredehöft, J.H., Cabane, M., Ehrenfreund, P., Gautier, T., Giri, C., Krüger, H., Le Roy, L., MacDermott, A.J. et al. (2015). COMETARY SCIENCE. Organic compounds on comet 67P/Churyumov-Gerasimenko revealed by COSAC mass spectrometry. *Science*, 349(6247), aab0689.

Goesmann, F., Brinckerhoff, W.B., Raulin, F., Goetz, W., Danell, R.M., Getty, S.A., Siljeström, S., Mißbach, H., Steininger, H., Arevalo, R.D. et al. (2017). The Mars Organic Molecule Analyzer (MOMA) instrument: Characterization of organic material in Martian sediments. *Astrobiology*, 17(6–7), 655–685.

Greenberg, J.M. (1984). Chemical evolution in space. *Origins of Life*, 14(1), 25–36.

Griesbeck, A.G. and Meierhenrich, U.J. (2002). Asymmetric photochemistry and photochirogenesis. *Angewandte Chemie International Edition*, 41(17), 3147–3154.

Guzman, M., Szopa, C., Freissinet, C., Buch, A., Stalport, F., Kaplan, D., Raulin, F. (2020). Testing the capabilities of the Mars Organic Molecule Analyser (MOMA) chromatographic columns for the separation of organic compounds on Mars. *Planetary and Space Science*, 186, 104903.

Hazen, R.M., Filley, T.R., Goodfriend, G.A. (2001). Selective adsorption of L- and D-amino acids on calcite: Implications for biochemical homochirality. *Proceedings of the National Academy of Sciences*, 98(10), 5487–5490.

Hegstrom, R.A. and Kondepudi, D.K. (1990). The handedness of the universe. *Scientific American*, 262(1), 108–115.

Hough, J.H. and Aitken, D.K. (2003). Polarimetry in the infrared: What can be learned? *Journal of Quantitative Spectroscopy and Radiative Transfer*, 79–80, 733–740.

Karagunis, G. and Coumoulos, G. (1938). A new method of resolving a racemic compound. *Nature*, 142(3586), 162–163.

Kondepudi, D.K. (1987). Selection of molecular chirality by extremely weak chiral interactions under far-from-equilibrium conditions. *Biosystems*, 20(1), 75–83.

Kondepudi, D.K. and Nelson, G.W. (1985). Weak neutral currents and the origin of biomolecular chirality. *Nature*, 314(6010), 438–441.

Kondepudi, D.K., Kaufman, R.J., Singh, N. (1990). Chiral symmetry breaking in sodium chlorate crystallization. *Science*, 250(4983), 975–976.

Lee, T.D. and Yang, C.N. (1956). Question of parity conservation in weak interactions. *Physical Review*, 104(1), 254–258.

Leseigneur, G., Filippi, J.-J., Baldovini, N., Meierhenrich, U. (2022a). Absolute configuration of aliphatic hydrocarbon enantiomers identified by gas chromatography: Theorized application for isoprenoid alkanes and the search of molecular biosignatures on Mars. *Symmetry*, 14(2), 326.

Leseigneur, G., Bredehöft, J.H., Gautier, T., Giri, C., Krüger, H., MacDermott, A.J., Meierhenrich, U.J., Muñoz Caro, G.M., Raulin, F., Steele, A. et al. (2022b). COSAC's only gas chromatogram taken on comet 67P/Churyumov-Gerasimenko. *ChemPlusChem*, 87(6), e202200116.

Leseigneur, G., Bredehöft, J.H., Gautier, T., Giri, C., Krüger, H., MacDermott, A.J., Meierhenrich, U.J., Muñoz Caro, G.M., Raulin, F., Steele, A. et al. (2022c). ESA's cometary mission Rosetta-Re-characterization of the COSAC mass spectrometry results. *Angewandte Chemie International Edition*, 61(29), e202201925.

Li, J., Shen, Z., Wang, J., Chen, X., Li, D., Wu, Y., Dong, J., Zhao, R., Gou, W., Wang, J. et al. (2017). Widespread presence of glycolaldehyde and ethylene glycol around Sagittarius B2. *The Astrophysical Journal*, 849(2), 115.

Lucas, P.W., Hough, J.H., Bailey, J., Chrysostomou, A., Gledhill, T.M., McCall, A. (2005). UV circular polarisation in star formation regions: The origin of homochirality? *Origins of Life and Evolution of Biospheres*, 35(1), 29–60.

MacDermott, A.J. (1995). Electro weak enantioselection and the origin of life. *Origins of Life and Evolution of the Biosphere*, 25(1–3), 191–199.

MacDermott, A.J., Barron, L.D., Brack, A., Buhse, T., Drake, A.F., Emery, R., Gottarelli, G., Greenberg, J.M., Haberle, R., Hegstrom, R.A. et al. (1996). Homochirality as the signature of life: The SETH cigar. *Planetary and Space Science*, 44(11), 1441–1446.

de Marcellus, P., Meinert, C., Nuevo, M., Filippi, J.-J., Danger, G., Deboffle, D., Nahon, L., Le Sergeant d'Hendecourt, L., Meierhenrich, U.J. (2011). Non-racemic amino acid production by ultraviolet irradiation of achiral interstellar ice analogs with circularly polarized light. *The Astrophysical Journal*, 727(2), L27.

Mason, S.F. (1984). Origins of biomolecular handedness. *Nature*, 311(5981), 19–23.

Mason, S.F. and Tranter, G.E. (1985). The electroweak origin of biomolecular handedness. *Proceedings of the Royal Society of London A. Mathematical and Physical Sciences*, 397(1812), 45–65.

Meierhenrich, U. (2008). *Amino Acids and the Asymmetry of Life: Caught in the Act of Formation*. Springer, Berlin/Heidelberg.

Meierhenrich, U. (ed.) (2014). *Comets and their Origin: The Tool to Decipher a Comet*. Wiley-VCH Verlag GmbH & Co. KGaA, Weinheim [Online]. Available at: http://doi.wiley.com/10.1002/9783527412778 [Accessed 20 July 2022].

Meierhenrich, U. and Thiemann, W.H.-P. (2004). Photochemical concepts on the origin of biomolecular asymmetry. *Origins of Life and Evolution of the Biosphere*, 34(1), 111–121.

Meinert, C., de Marcellus, P., Le Sergeant d'Hendecourt, L., Nahon, L., Jones, N.C., Hoffmann, S.V., Bredehöft, J.H., Meierhenrich, U.J. (2011). Photochirogenesis: Photochemical models on the absolute asymmetric formation of amino acids in interstellar space. *Physics of Life Reviews*, 8(3), 307–330.

Meinert, C., Myrgorodska, I., de Marcellus, P., Buhse, T., Nahon, L., Hoffmann, S.V., d'Hendecourt, L.L.S., Meierhenrich, U.J. (2016). Ribose and related sugars from ultraviolet irradiation of interstellar ice analogs. *Science*, 352(6282), 208–212.

Meinert, C., Garcia, A.D., Topin, J., Jones, N.C., Diekmann, M., Berger, R., Nahon, L., Hoffmann, S.V., Meierhenrich, U.J. (2022). Amino acid gas phase circular dichroism and implications for the origin of biomolecular asymmetry. *Nature Communications*, 13(1), 502.

Modica, P., Meinert, C., de Marcellus, P., Nahon, L., Meierhenrich, U.J., d'Hendecourt, L.L.S. (2014). Enantiomeric excesses induced in amino acids by ultraviolet circularly polarized light irradiation of extraterrestrial ice analogs: A possible source of asymmetry for prebiotic chemistry. *The Astrophysical Journal*, 788(1), 79.

Muñoz Caro, G.M., Meierhenrich, U.J., Schutte, W.A., Barbier, B., Arcones Segovia, A., Rosenbauer, H., Thiemann, W.H.-P., Brack, A., Greenberg, J.M. (2002). Amino acids from ultraviolet irradiation of interstellar ice analogues. *Nature*, 416(6879), 403–406.

Myrgorodska, I., Meinert, C., Martins, Z., Le Sergeant d'Hendecourt, L., Meierhenrich, U.J. (2015). Molecular chirality in meteorites and interstellar ices, and the chirality experiment on board the ESA cometary Rosetta mission. *Angewandte Chemie International Edition*, 54(5), 1402–1412.

Pizzarello, S., Zolensky, M., Turk, K.A. (2003). Nonracemic isovaline in the Murchison meteorite: Chiral distribution and mineral association. *Geochimica et Cosmochimica Acta*, 67(8), 1589–1595.

Ruiz-Mirazo, K., Briones, C., de la Escosura, A. (2014). Prebiotic systems chemistry: New perspectives for the origins of life. *Chemical Reviews*, 114(1), 285–366.

Siegel, J.S. (1998). Homochiral imperative of molecular evolution. *Chirality*, 10(1–2), 24–27.

Soai, K., Shibata, T., Morioka, H., Choji, K. (1995). Asymmetric autocatalysis and amplification of enantiomeric excess of a chiral molecule. *Nature*, 378(6559), 767–768.

Thiemann, W.H.-P. and Meierhenrich, U. (2001). ESA mission ROSETTA will probe for chirality of cometary amino acids. *Origins of Life and Evolution of the Biosphere*, 31(1), 199–210.

Tverdislov, V.A., Yakovenko, L.V., Zhavoronkov, A.A. (2007). Chirality as a problem of biochemical physics. *Russian Journal of General Chemistry*, 77(11), 1994–2005.

Tverdislov, V.A., Malyshko, E.V., Il'chenko, S.A., Zhulyabina, O.A., Yakovenko, L.V. (2017). A periodic system of chiral structures in molecular biology. *Biophysics*, 62(3), 331–341.

Ulamec, S., Fantinati, C., Maibaum, M., Geurts, K., Biele, J., Jansen, S., Küchemann, O., Cozzoni, B., Finke, F., Lommatsch, V. et al. (2016). Rosetta Lander – Landing and operations on comet 67P/Churyumov–Gerasimenko. *Acta Astronautica*, 125, 80–91.

Uomini, N.T. (2009). The prehistory of handedness: Archaeological data and comparative ethology. *Journal of Human Evolution*, 57(4), 411–419.

Vago, J.L., Westall, F., Coates, A.J., Jaumann, R., Korablev, O., Ciarletti, V., Mitrofanov, I., Josset, J.-L., De Sanctis, M.C., Bibring, J.-P. et al. (2017). Habitability on early Mars and the search for biosignatures with the ExoMars rover. *Astrobiology*, 17(6–7), 471–510.

Weissbuch, I., Leiserowitz, L., Lahav, M. (2005). Stochastic "mirror symmetry breaking" via self-assembly, reactivity and amplification of chirality: Relevance to abiotic conditions. In *Prebiotic Chemistry*, Walde, P. (ed.). Springer, Berlin [Online]. Available at: https://doi.org/10.1007/b137067 [Accessed 14 December 2022].

Wu, C.S., Ambler, E., Hayward, R.W., Hoppes, D.D., Hudson, R.P. (1957). Experimental test of parity conservation in beta decay. *Physical Review*, 105(4), 1413–1415.

Zepik, H.H., Rajamani, S., Maurel, M.-C., Deamer, D. (2007). Oligomerization of thioglutamic acid: Encapsulated reactions and lipid catalysis. *Origins of Life and Evolution of Biospheres*, 37(6), 495–505.

3

The Role of Formamide in Prebiotic Chemistry

Raffaele SALADINO[1], Giovanna COSTANZO[2]
and Bruno Mattia BIZZARRI[1]
[1] Ecological and Biological Sciences Department, University of Tuscia, Viterbo, Italy
[2] Institute of Molecular Biology and Pathology, CNR, Rome, Italy

3.1. Introduction

One of the most humbling aspects in prebiotic chemistry is the fact that we know remarkably little about the geochemical and geophysical conditions of the Earth at the origin of life. The pivotal Miller-Urey experiment suggested a possible path to set a first model of molecular evolution, that is energy (electric discharge) and matter (primitive atmosphere) combined together to yield chemical complexity (Miller 1953). In this latter case (and in the successive modifications of the original set), amino acids and hydroxy acids were synthesized as possible building blocks for the formation of peptides and possibly proto-proteins. A large panel of organic molecules of biological relevance, including the complete set of RNA's nucleobases, was successively synthesized by more and more experiments exploring gas mixtures of different compositions and a variety of energy sources (Miyakawa et al. 2002a; Johnson et al. 2008; Cooper et al. 2017). The first significant clue coming from these studies was the comprehension that chemical complexity and scaffold variability could be generated from a limited number of reagents, and among them, hydrogen cyanide (HCN) caught the attention of the scientists (Ferris et al. 1972). HCN was hypothesized to be involved in a complex network of chemical transformations during the electric discharge, including the Strecker condensation of imines to amino acids (Parker et al. 2011), as well as self-oligomerization to purines

The First Steps of Life,
coordinated by Ernesto DI MAURO. © ISTE Ltd 2024.

(Orò 1963). Some years after, experimental and computational data showed that formamide (NH_2CHO) was involved in the formation of amino acids (Saitta and Saija 2014) and nucleic acid bases (Ferus et al. 2017) under Miller-Urey like conditions (Ferus et al. 2015). Formamide is a well-recognized chemical precursor in prebiotic chemistry (Saladino et al. 2012a). It has been recognized in a large variety of low and intermediate-mass pre-stellar and proto-stellar objects (López-Sepulcre et al. 2015), including Sagittarius B2 (Rubin et al. 1971), massive hot molecular cores (Bisschop et al. 2007; Adande et al. 2011), low-mass proto-stellar IRAS 16293−2422 (Kahane et al. 2013), and outflow shock regions L1157-B1/B2 (Mendoza et al. 2014). The detection of NH_2CHO in comet Hale-Bopp (Bockelée-Morvan et al. 2000), C/2012 F6 and C/2013 R1 (Biver et al. 2014), and 67P/Churyumov-Gerasimenko (Goesmann et al. 2015) was also reported. The ubiquitous presence of NH_2CHO in the Cosmos is associated with gas-phase (Barone et al. 2015) and grain-surface chemistry (Fedoseev et al. 2015), as well as shocks produced by fast proto-stellar jets propagating through high-dense gas surrounding the newly born star regions (Lefloch et al. 2017). In addition, the presence of liquid NH_2CHO below the frozen surface of the mantle on Titan and the Jupiter's satellite Europa has been reported (Levy et al. 2000; Parnell et al. 2006). Astronomical sources of NH_2CHO (Lòpez-Sepulcre 2019) and details on the reactivity of this compound in space conditions have been reviewed and discussed (Carota et al. 2015). At the planetary level (and in the habitable zone) where the presence of water is expected, the prebiotic chemistry of NH_2CHO is connected to that of HCN (Costanzo et al. 2007a). HCN provides two main reaction pathways, after the solubilization in water, which are as follows: (i) self-oligomerization to dimer (AMN) and trimer (DAMN) or (ii) hydrolysis to yield NH_2CHO. While the two opposite reactions work with a similar kinetic with a (relatively) high concentration of HCN (0.01–0.1 M), the formation of NH_2CHO largely prevails in alkaline dilute solutions (Sanchez et al. 1967). Theoretical studies demonstrated that the concentration of HCN in the primitive ocean was far too low for the occurrence of self-oligomerization, favoring NH_2CHO (Stribling and Miller 1987; Miyakawa et al. 2002b). Even if NH_2CHO can be in turn hydrolyzed to ammonium formate (half-life of ca 200 years at 25°C and pH 7.0; Slebocka-Tilk et al. 2002), it can be concentrated (at difference of HCN) in a drying lagoon environment under a large variety of geochemical conditions (Saladino et al. 2012b). In addition, the concentration of NH_2CHO in local environments can be increased by adsorption in minerals, such as clays (Joussein et al. 2007), silicates (Signorile et al. 2018) and NaCl crystals (Pastero et al. 2012), which are all representative of plausible geochemical scenarios in the primitive Earth (Bizzarri et al. 2021a). Additional pathways for the synthesis of NH_2CHO include the reaction of ammonia with formic acid and formic acid esters (Lei et al. 2010; Rimola et al. 2005), sparking discharge in neutral-like atmosphere (Abelson 1956), radioactive mineral transformations

(Adam et al. 2018), ammonium formate dehydratation (Sponer et al. 2016a) and high energy plasma impact events (Ferus et al. 2014). Overall, these results confirm NH_2CHO as a robust chemical precursor in prebiotic chemistry (Pino et al. 2015).

3.2. Effect of minerals and self-organization in the prebiotic chemistry of formamide

3.2.1. Surface catalysis and geochemical scenarios

Purine and adenine (one of RNA and DNA nucleic acid bases) were synthesized for the first time from NH_2CHO by Yamada et al. (1978a). They reported that two equivalents of NH_2CHO and three equivalents of HCN (produced in situ by NH_2CHO dehydration) combined together at 160°C to yield the purine derivatives (Yamada et al. 1978b). The reaction pathway involved the formation of a dimer, followed by the elimination of water to afford N-formyl-formamidine and the successive addition of HCN and NH_3 on this intermediate. Thus, NH_2CHO was degraded in situ (alternatively to possible self-oligomerization) to yield the reactive species required for the process. Examples of degradation of NH_2CHO to HCN (Cataldo et al. 2009), CO and NH_3 (Kakumoto et al. 1985), isocyanic acid (HNCO) (Buffaerts et al. 2019), ammonium formate ($NH_4^+COO^-$) and formic acid (HCOOH) (Gorb et al. 2005) have been reported. The selectivity and kinetics of these degradative processes are controlled by the presence of ice, as well as minerals and metal oxides. The adsorption of NH_2CHO at the surface of low-density amorphous ices (which are representative of the interstellar medium) produced multimolecular layers instead of the dissolution of the molecule in the bulk of the ice. The strong hydrogen bond interactions acting between the adsorbed NH_2CHO molecules, associated with their high lateral density, are then responsible for the enhancement of the reactivity of the system (Kiss et al. 2019). In a similar way, isolated silanol groups on the surface of silica improved the decomposition of NH_2CHO to HCN in the formation of nucleic acid bases (Rimola et al. 2013). The catalytic effect played by silica and simple metal oxides in the prebiotic chemistry of NH_2CHO was reported by Saladino et al. (2001) during the one-pot synthesis of adenine, cytosine and 4-hydroxypyrimidine [4(3H)-pyrimidinone]. The analysis of the reaction pathway confirmed the multifunctional role of NH_2CHO, which was involved in the formation of a dimer, followed by disproportionation and successive addition/elimination of NH_3 (Saladino et al. 2007). The dependence of silica solubility from pH further enlarged the prebiotic chemistry of formamide. At alkaline conditions (pH from 9.5 to 11.5), the precipitation of silica is associated with the formation of self-assembled nanocrystalline materials (Kellermeier et al. 2013), which are classified as silica/carbonate biomorphs (García-Ruiz et al. 2003).

The further increase of the pH (12) produces chemical gardens consisting of tubular structures of metal–[hydroxide/silica(te)] bearing an inorganic membrane that is able to separate solutions with different compositions (Glaab et al. 2017). The outer part of the membrane is rich in amorphous silica, while the inner part is made by nanocrystalline and microcrystalline metal oxyhydroxides (Glaab et al. 2016). Both silica/carbonate biomorphs and chemical gardens are reminiscent geochemical structures of the serpentinization process occurring on the primitive Earth (Neubeck et al. 2011). Interestingly, chemical gardens made from pellets of soluble metallic salts are able to catalyze the condensation of NH_2CHO in saturated sodium silicate solutions (pH 12) to yield nucleic acid bases, amino acids and carboxylic acids (Saladino et al. 2016a). The collections of products synthesized inside and outside the membrane were site-specific, nucleic acid bases prevailing inside and carboxylic acids outside of the membrane, as a consequence of the effect played by the mineral composition of the membrane on the selectivity of the transformation. Note that the membrane provided space compartmentalization and catalysis, increasing the spatial control of the system in the synthesis of chemical precursors of pre-metabolic and pre-genetic apparatuses (Saladino et al. 2019). Similar results were obtained using silica microvesicles. In this latter case, the condensation of NH_2CHO to biomolecules occurred during the time at which the metal/silicate membrane formed at the interface between the metallic solution and the surrounding alkaline silicate solution (Bizzarri et al. 2018). Irrespective from the specific type of silica self-organized structures, the synthesis of biomolecules was very efficient in terms of both yield and scaffold variability, especially in the presence of common components of the ultramafic and komaitic rocks which are characteristic of the earliest crust of the planet (Foustoukos and Seyfried 2004). Mineral vesicles were catalysts more efficient than chemical gardens, probably due to the low degree of crystallinity showed by their inorganic membrane. In a similar way, amorphous silicates mimicking cosmic dusts showed a reactivity higher than the crystalline counterpart in the thermal condensation of NH_2CHO (Saladino et al. 2005b). In analogy with chemical gardens and silica microvesicles, hydrothermal precipitates forming chimney-like structures are able to absorb chemical precursors (Russel et al. 2014). For this reason, submarine hydrothermal vents containing a suite of minerals such as iron/nickel sulfides and double-layer oxyhydroxides have been pointed out as important geochemical reactors for chemical evolution on Earth (Baaske et al. 2007), and their role in prebiotic chemistry has been reviewed (Sojo et al. 2016). The combination of thermophoresis and convection inside the hydrothermal pores of these geochemical structures leads to the accumulation of NH_2CHO up to the concentration useful for the synthesis of organic compounds (Niether et al. 2016). In addition, the thermal condensation of NH_2CHO in the presence of typical components of hydrothermal pores, such as sulfur containing minerals pyrrhotine ($Fe1-xS$), pyrite (FeS_2), chalcopyrite ($FeCuS_2$), bornite ($FeCu_5S_4$), and covellite

(CuS), was reported to yield a large panel of nucleic acid bases (Saladino et al. 2001, 2009). These minerals, which are representative components of the uppermost sections of conjugate chimney material (James et al. 2014), catalyzed the synthesis of cytosine, isocytosine (which is a guanine bioisoster) and adenine in appreciable yield (Saladino et al. 2008). In a similar way, the condensation of NH_2CHO in the presence of montmorillonites with different pore size distribution, surface area, cation exchange capability and acidity afforded adenine, cytosine, and uracil (Saladino et al. 2004). Along with these products, aminoimidazole-4-carboxamide (AICA), 5-formamidoimidazole-4-carboxamide and hypoxanthine, which are present-day intermediates of the inosine pathway in the cell, as well as N9-formylpurine containing a masked glycosidic bond, were isolated from the reaction mixture. The mechanism of formation of nucleic acid bases and precursors from NH_2CHO under thermal conditions has been studied and discussed in detail (Hudson et al. 2012; Šponer et al. 2012).

3.2.2. Chemomimesis, circularity and thermodynamic niches

Experimental evidence showed that minerals may have played a key role in the synthesis of molecular templates for modern biology. We mentioned previously that the intermediates of the present-day biosynthesis of inosine are synthesized in a one-pot way from NH_2CHO and montmorillonites. This is an example of chemomimesis, which is the occurrence of selective and reproducible prebiotic processes able to provide molecular templates for the aggregation of complex pre-biological pathways (Eschenmoser and Loewenthal 1992; Saladino et al. 2018a). In a similar way, thymine is synthesized from NH_2CHO in the presence of titanium dioxide (TiO_2), along with 5-hydroxymethyl uracil (5-HMU) and other pyrimidine and purine nucleic acid analogues (Saladino et al. 2003). The synthesis of thymine from NH_2CHO and TiO_2 was also reported under photochemical irradiation under single crystal-dry conditions (Senanayake and Idriss 2006). Titanium dioxide accounts for an appreciable part of the Earth's crustal and mantle rocks in the mineralogical form of perovskite (Meinhold 2010). The presence of 5-HMU suggested that thymine derived from the electrophilic addition of in situ generated formaldehyde at the C(5)-position of uracil, followed by the inter-molecular formic acid hydride-shift mechanism. This reaction pathway is a reminder of the present-day biosynthesis of thymidine in the cell, during which a formaldehyde-like residue is transferred from 5,10-methylene tetrahydrofolate to the C(5)-position of the uracil ring. In addition, different carboxylic acids, which are intermediates of the tri-carboxylic acid cycle (Krebs cycle), are synthesized from NH_2CHO and TiO_2 under UV irradiation (Saladino et al. 2011a) or, alternatively, under thermal condensation of NH_2CHO in the presence of borates (Saladino et al. 2011b) and zirconium minerals (Saladino et al. 2010). Thus, several ingredients of the

Krebs cycle ("the core of the core" of the metabolism; Nelson and Cox 2000) are synthesized simply by mixing NH_2CHO and minerals. NH_2CHO and minerals are also able to generate thermodynamic niches in which new synthesized biomolecules are stabilized with respect to possible degradation processes. In order to introduce this topic, both the conditions that control the synthesis of biomolecules from NH_2CHO and those leading to their degradation must be determined. This coupled approach is necessary since the stability of the organic molecules is a major concern in the geochemical scenarios dealing with the origin of life (Levy and Miller 1998; Bada and Lazcano 2002). The reaction of oligonucleotides with NH_2CHO during one-lane chemical DNA sequencing procedures occurs by two reactive steps, the degradation of purine and pyrimidine nucleic acid bases followed by beta-elimination at the sugar moiety (Di Mauro et al. 1994). In the first step, NH_2CHO performs a nucleophilic addition at the C-8 position of the purine, followed by the ring opening of the imidazole ring and the formation of formamido-pyrimidine nucleosides (FAPyr). The successive hydrolysis of the N-glycosidic bond yields the sugar and the corresponding formamido-pyrimidine as a living group (Saladino et al. 1996). In the case of pyrimidine nucleic acid bases, the contemporary addition of NH_2CHO at the C-6 positions of the pyrimidine ring is followed by the release of NH_2CHO and the formation of low molecular weight fragments, such as urea, ammonia and beta-dicarbonyl derivatives (Negri et al. 1996; Saladino et al. 1997). Given that purines embedded into oligonucleotides react with NH_2CHO to yield pyrimidines, and pyrimidines react with NH_2CHO to yield precursors for the synthesis of both purines and pyrimidines (Saladino et al. 2001), the problem of the stability of the latter compounds in the primitive Earth (Shapiro 1995, 1996) is overcome by the circularity of the chemistry of NH_2CHO (Šponer et al. 2016b) (Figure 3.1).

In addition, NH_2CHO acts as an organo-catalyst. In fact, it is regenerated during the overall transformation (Closs et al. 2020). The circularity of NH_2CHO has recently been revisited in the synthesis of purine RNA ribonucleotides from FAPyr intermediates (Becker et al. 2016, 2019; Saladino et al. 2018b). Associated with circularity, the existence of thermodynamic niches favorable for the stability of biopolymers further improves the relevance of NH_2CHO in prebiotic chemistry. Pioneering studies revealed that the polymeric state of phosphorylated nucleosides is thermodynamically stabilized by NH_2CHO with respect to the monomeric one, even at high temperature, generating defined thermodynamic niches inside which RNA (or, in alternative, DNA) are the most stable structures (Saladino et al. 2006a; Soukup and Breaker 1999). The stability of RNA in the H_2O/NH_2CHO mixture is higher than that of the simple ribonucleotide in the range of 33%–100 % v/v at 60°C–90°C (Saladino et al. 2005). The borderline of these thermodynamic niches is controlled by the presence of minerals. Borates, such as ludwigite $Mg_2Fe^{3+}BO_5$, hambergite $Be_2(BO_3)OH$ and jeremejevite $Al_6B_5O_{15}F_{2.5}(OH)0.5$, showed the highest protective effect in the stabilization of RNA (Cossetti et al. 2010), in accordance with the well-established protection exerted by borates on ribose (Ricardo et al. 2004; Furukawa et al. 2013). In addition, borates favored the regiospecific phosphorylation of ribonucleosides with activated inorganic phosphates (Kim et al. 2016). Copper containing minerals libethenite $Cu^{2+}_2(PO_4)OH$, cornetite $Cu^{2+}_3(PO_4)(OH)_3$ and malachite $Cu^{2+}_5(PO_4)_2(OH)_4$ also protected RNA from degradation (Saladino et al. 2006b). In addition, they were able to catalyze the phosphorylation of nucleosides to nucleotides due to the release of free phosphate from the bulk of the mineral (Costanzo et al. 2007b). On the contrary, zirconium (Saladino et al. 2010) and Fe–S–Cu minerals (Saladino et al. 2008), and to a lower extent, Murchison materials (Saladino et al. 2011c), increased RNA degradation. In these latter cases, the futile cycle of synthesis and degradation of informational polymers may be solved by the circularity of prebiotic chemistry of NH_2CHO, and the stability of nucleic acid bases is not per se a major requirement for the occurrence of molecular evolution.

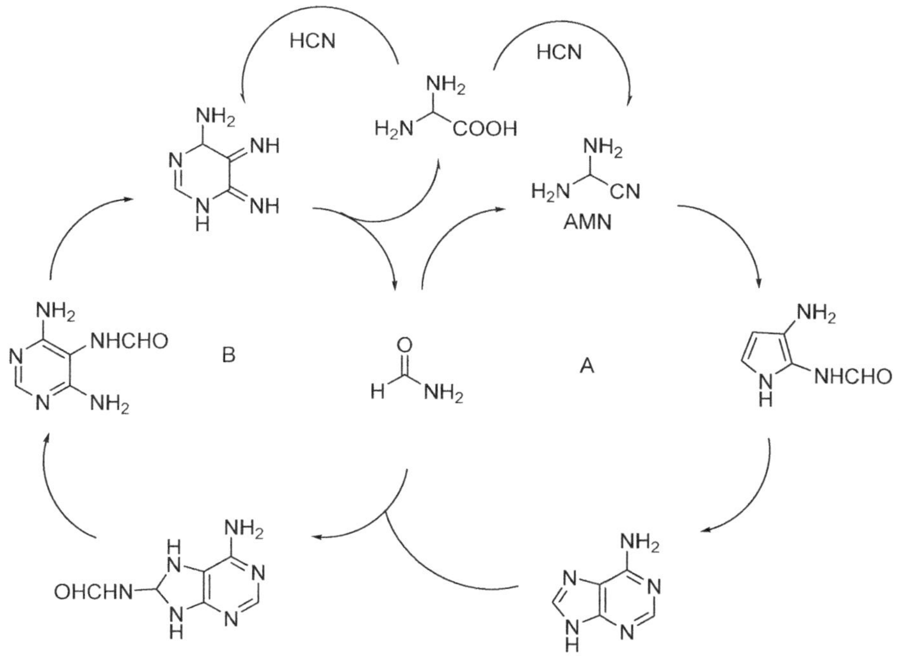

Figure 3.1. Circularity in the prebiotic chemistry of NH_2CHO

NOTE ON FIGURE 3.1.– *Panel A: synthesis of adenine from NH_2CHO. Panel B: degradation of adenine by NH_2CHO with formation of pyrimidine, followed by* *degradation of pyrimidine and production of AMN and NH$_2$CHO again. The synthetic and degradative cycles represented in Panels A and B are connected together by NH$_2$CHO and HCN. The regeneration of part of the NH$_2$CHO at the end of the degradative process highlights its role as an organo-catalyst.*

3.2.3. Nucleosides phosphorylation

Minerals also played a role in the phosphorylation of nucleosides to nucleotides. A general problem derives from the fact that the formation of a phosphodiester bond is thermodynamically uphill. Thus, the template-directed protein-free prebiotic synthesis of phosphodiester-linked oligonucleotides most probably required the use of chemically activated nucleotides. A recent review about prebiotic nucleosides phosphorylation is given in Guo et al. (2023). Moving on from the seminal observations by Schoffstall, the effect of formamide on the phosphorylation of nucleosides by using both inorganic phosphates and nucleoside mono-phosphates (NMPs) as phosphate donors, was investigated. Formamide is an efficient promoter of the reaction and phosphorylated nucleosides may amount, at the tested temperature of 90°C, to up to ca. 20% of the nucleoside input. Phosphorylation occurred in pure formamide on adenosine or cytidine (the two analyzed nucleosides), with similar efficiency, regardless of the source of phosphate, be it soluble KH$_2$PO$_4$ or an MNP. In this latter case, transphosphorylation showed an initial lag, corresponding to the time needed to release the phosphate moiety from the donor-phosphorylated nucleoside. The acceptor nucleoside was alternately phosphorylated in 5'-, 3'-, 2'-, or 2', 3' and 3', 5'-cyclic positions. The form appearing first, and amounting to higher levels, was adenosine 5'-monophosphate (5'-AMP), followed by the 3'-, the 2'-, and the cyclic forms. Due to the different half-lives of the phosphorylated nucleosides (cyclic > 5' > 2' > 3'), and the efficiently occurring internucleotide transphosphorylation, the pool of the phosphorylated nucleosides progressively enriched with the 2',3' and 3',5'-cyclic forms (Saladino et al. 2009). Efficient phosphorylation of nucleosides was also observed in formamide using numerous phosphate minerals as phosphate donors (Costanzo et al. 2007b). A group of minerals were identified to have a very active ability to promote phosphorylation: libethenite Cu$^{2+}_2$(PO$_4$)(OH), ludjibaite Cu$^{2+}_5$(PO$_4$)$_2$(OH)$_4$, reichenbachite Cu$^{2+}_5$(PO$_4$)$_2$(OH)$_4$, cornetite Cu$^{2+}_3$(PO)(OH)$_3$ and hydroxylapatite Ca$_5$(PO$_4$)$_3$OH. With the exception of hydroxylapatite, these are all copper phosphate minerals and are chemically very similar (in one case identical). They do at the same time release phosphate and copper ions that could act as catalysts in the phosphorylation reaction. In addition, borates favored the regiospecific phosphorylation of ribonucleosides with activated inorganic phosphates (Kim et al. 2016).

3.3. Continuity and mineral complexity

In the previous paragraphs, we described the effect played by minerals in the synthesis of nucleic acid bases, amino acids and carboxylic acids. Now, in the context of molecular Darwinism, we will correlate the continuity between the simple building blocks and complex (pre)genetic and (pre)metabolic biopolymers. The major gap to reach this goal is the difficulty for a synthetic pathway able to increase the level of chemical complexity (e.g. nucleosides, nucleotides and oligonucleotides) in the same experimental context in which the building blocks are produced. How can the mineral catalyst affect (and favor) this result? A clear-cut case is that of meteorites. Studies over the past five decades showed that meteorites are crucial ingredients for the origin of life on Earth and possibly elsewhere in the Universe (Burton et al. 2018; Martin 2018).

They are very complex aggregates of different minerals and metal oxides (in both crystalline and amorphous states) which have been modified during a long evolutionary process, starting from the original sky body (Ruf et al. 2018). Meteorites work as efficient catalysts for the synthesis of biomolecules from NH_2CHO under thermal conditions. Twelve meteorite specimens, representative of their major classes (iron, stony iron, achondrites and chondrites), catalyzed the synthesis of 13 nucleic acid bases and analogues (including uracil, cytosine, adenine and guanine), 17 carboxylic acids (comprising five intermediates of the Krebs cycle and two intermediates of glycolysis and gluconeogenesis), 13 amino acids (seven of which were proteogenic) and a miscellanea of low molecular weight condensing agents (guanidine, DAMN, urea and carbodiimide) (Saladino et al. 2013). Similar results were obtained from NH_2CHO and meteorites of the chondrite type (CV3, CR2, CO3, CK4, C-ung and C-2ung) in the presence of thermal water (spring pool, Bagnaccio, Viterbo, Italy), seawater (Mediterranean area, Montalto di Castro, Viterbo, Italy) and distilled water (Rotelli et al. 2016). Thermal water and, to a lower extent, seawater, were more reactive than distilled water, the selectivity of the reactions being tuned by the composition of each chondrite (reactivity order: CR> CV3 (CK3)> CO3). These results have been extended to the case of Phlegrean Hydrothermal Fields (Naples Italy) (Botta et al. 2018), confirming that, unlike theoretical argumentation of the contrary (Bada et al. 2016), NH_2CHO/water is a flexible and fertile prebiotic incubator. Note that the reactivity of meteorites was higher than that of their individual components due to the occurrence of the synergistic effect between the minerals included in the meteorite.

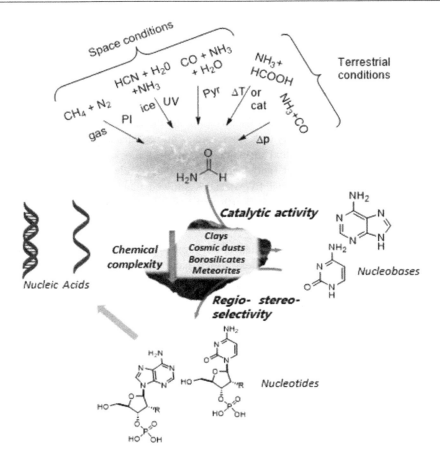

Figure 3.2. NH_2CHO is easily produced from a variety of planetary and space conditions. Mineral complexity favors the increasing of chemical complexity in the biomolecules obtained from NH_2CHO. Simple minerals and metal oxides catalyze the synthesis of nucleic acid bases, while nucleosides and nucleotides are obtained in the presence of meteorites. For a color version of this figure, see www.iste.co.uk/dimauro/firststeps.zip

Thus, meteorites worked as chemical reactors for prebiotic chemistry in addition to their traditional role of carrier of organic compounds during their permanence in space (Chyba 1987; Burton et al. 2012). In addition, the very fact that the syntheses of biomolecules may also occur in meteorites during their non-terrestrial history, points out that these processes are not exclusively terrestrial.

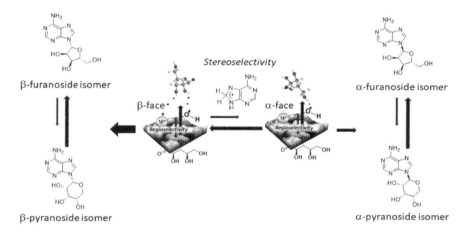

Figure 3.3. *Regioselective and stereoselective synthesis of adenosine from NH₂CHO and meteorites. For a color version of this figure, see www.iste.co.uk/dimauro/firststeps.zip*

NOTE ON FIGURE 3.3.– *Regioselectivity was controlled by the preferential cyclization of the sugar to the five-membered furanose ring when adsorbed on the surface of the meteorite. Stereoselectivity derived from the selective OH-geminal interaction of the cyclized sugar with the surface of the meteorite, followed by the addition of the radical from the beta-site of the mineral-sugar complex.*

Even more relevant results have been obtained by irradiation of NH_2CHO and meteorites with high energy proton beams mimicking the solar wind. In this latter case, four nucleosides bearing the beta-natural linkage between the nucleic acid base and sugar (uridine, cytidine, adenosine and thymidine) were obtained from NH_2CHO in a one-pot way, in addition to 14 nucleic acid bases and analogues (including the complete set of nucleic acid bases in both RNA and DNA), 22 carboxylic acids (including six intermediates of the Krebs cycle, two intermediates of glycolysis and gluconeogenesis and one intermediate of the Cori cycle), nine amino acids and eight sugars (including ribose and 2′-deoxyribose) (Saladino et al. 2015) (Figure 3.3). The one-pot synthesis of nucleosides from NH_2CHO is of particular interest in the context of prebiotic chemistry, since previous studies were mainly focused on multi-steps procedures based on the construction of the nucleic acid base on functionalized sugar (Powner et al. 2009), and phosphorylation of anhydronucleoside intermediates (Powner et al. 2010). Nucleic acid bases, carboxylic acids and amino acids were also obtained by the irradiation of NH_2CHO and meteorites with high-energy heavy ions, mimicking the effect of Cosmic radiation (Saladino et al. 2016b). The high regioselectivity and stereoselectivity obtained in these syntheses was interpreted as a consequence of two concomitant processes, the thermodynamic stability of centered radical species and the favorable absorption of the sugar on the surface of the meteorite prior to the addition of the radical (Saladino et al. 2017). In the case of regioselectivity, the most stable product in the radiolysis of purine is the C8-hydrogenated adduct (Wetmore et al. 1988). The coupling of this radical with the sugar preferentially occurred at the (natural) C(9)-position of the purine ring with respect to alternative C(7) and N(6)-positions, due to the possibility of the elimination of a water molecule in an exothermic reaction step. In the stereoselectivity of the process, the meteorite was able to stabilize the sugar in the (natural) furanose anomeric form (Lambert et al. 2004; Georgelin et al. 2015), favoring the addition of the radical at the less hindered face of the sugar moiety, which was opposite to the geminal OH groups involved in the interaction with the mineral surface (beta-position) (Figure 3.3). As an extension of this study, di-glycosylated adenines, which are "remnant" structural motifs in damaged DNA sites (Price et al. 2014), have been applied in the one-pot glycosylation of pyrimidine nucleic acid bases under proton beam irradiation. In this latter case, the reaction proceeded by selective intermolecular transfer of the N6-glycosyl moiety to the nucleic acid base acceptor (Bizzarri et al. 2021b). The transfer of the glycosyl moiety was highly regioselective to yield N1-pyrimidine nucleosides, with the natural β-anomers prevailing in the presence of NH_2CHO and chondrite meteorites, especially in the case of meteorites characterized by Lewis acid transition metal sites usually active in the glycosylation of carbohydrates (Pfaffe and Mahrwald 2012). This reaction is the first example of a convergent procedure for the contemporary synthesis of purine and pyrimidine nucleosides (Biscans 2018), since the complete set of nucleosides was obtained from only one di-glycosylated derivative when the appropriate nucleic acid base was available as a glycosyl acceptor. The pair-wise scatter plots and Pearson correlation analysis (Emerson et al. 2013) of the product-to-product relationships of the reaction between NH_2CHO and 19 meteorites under different energy conditions (thermal condensation, thermal condensation in NH_2CHO/H_2O mixture, and high energy proton beam irradiation) showed a high value of the correlation probability in the contemporary synthesis of nucleic acid bases, carboxylic acids and amino acids. Thus, irrespective of the type of meteorite (and the specific energy source), the three major classes of biomolecules were always synthesized from NH_2CHO (and NH_2CHO/H_2O mixture) and meteorites. High molecular weight carboxylic acids were not correlated; alternatively, they were negatively correlated to the other types of organic compounds (Bizzarri et al. 2021).

3.4. Energy-driven selectivity

Energy sources affect the chemoselectivity and regioselectivity of the synthesis of nucleic acid bases from NH_2CHO. As a representative example, the thermal condensation of NH_2CHO in the presence of calcium carbonate yields purine as the only recovered product (Saladino et al. 2001). On the contrary, when the reaction was repeated under photothermal conditions (that is, the contemporaneous heating and UV-irradiation of the reaction mixture), the formation of a large set of nucleic acid bases, including adenine, hypoxanthine and guanine, was obtained in appreciable yield (Barks 2010). The improved efficacy of the prebiotic process observed in the latter case was associated with the selective production of reactive HCN oligomers (Zubay and Mui 2001; Orgel 2004) as a consequence of the photochemical degradation of NH_2CHO.

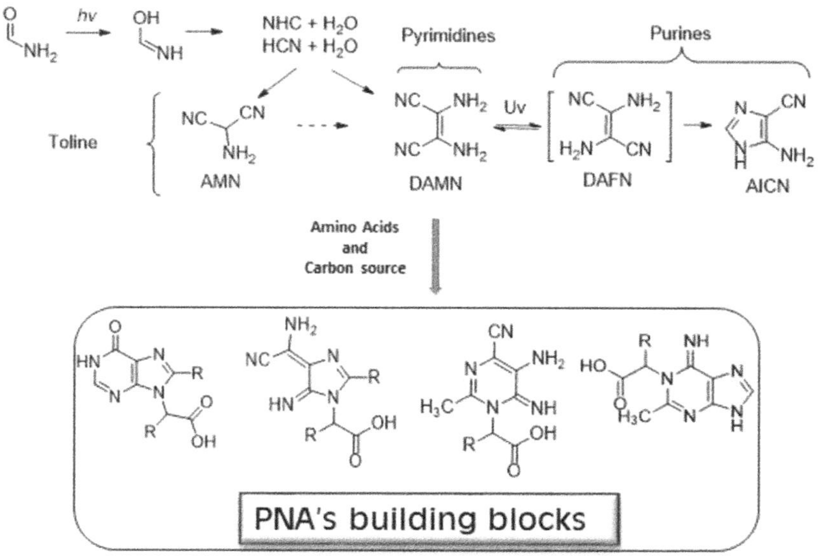

Figure 3.4. *Multi-way regioselective synthesis of amino acid decorated imidazole, purine and pyrimidine derivatives resembling PNA's building blocks by multicomponent chemistry starting from prebiotic DMAN and AMN*

Formamide is a very effective photoactive entity, and its involvement in the UV-mediated synthesis of nucleic acid bases and HCN oligomers, including diaminomaleonitrile (DAMN), diaminofumaronitrile (DAFN), aminoimidazolecarbonitrile (AICN) and aminoimidazolecarboxamide (AICA), has recently been reported (Di Mauro et al. 2021). In addition, the formation of DAMN (and of aminomalonitrile AMN) as key intermediates in the prebiotic chemistry of NH_2CHO can control the efficacy of the synthesis of nucleic acid bases and analogues. In this latter case, energy sources control the selectivity of the process (that is, thermal versus photochemical activation), as recently discussed and analyzed in the context of prebiotic multicomponent chemistry (Bizzarri et al. 2022a). What is evident from these studies is that the formation of the HCN intermediates is under the control of the energy source, and that different HCN intermediates may afford heterocycles characterized by rings of different sizes and chemical compositions. Thus, while imidazole derivatives prevail under thermal conditions by a direct transformation of DAMN, photochemical and photothermal conditions shifted the selectivity of the reaction toward the formation of DAFN and AICN intermediates, followed by selective synthesis of pyrimidine and purine derivatives resembling peptide nucleic acids (PNA) building blocks, respectively (Bizzarri et al. 2022b; Bizzarri et al. 2021c) (Figure 3.4). This result further connects the prebiotic chemistry of NH_2CHO to the synthesis of alternative RNA molecules.

3.5. References

Abelson, P.H. (1956). Amino acids formed in primitive atmospheres. *Science*, 124, 935.

Adam, Z.R., Hongo, Y., Cleaves II, H.J., Yi, R., Fahrenbach, A.C., Yoda, I., Aono, M. (2018). Estimating the capacity for production of formamide by radioactive minerals on the prebiotic Earth. *Sci. Rep.*, 8, 265.

Adande, G.R., Woolf, N.J.G., Ziurys L.M. (2011). Observations of interstellar formamide: Availability of a prebiotic precursor in the galactic habitable zone. *Astrobiology*, 13, 439.

Baaske, P., Weinert, F.M., Duhr, S., Lemke, K.H., Russel, M.J., Braun, D. (2007). Extreme accumulation of nucleotides in simulated hydrothermal pore systems. *Proc. Natl. Acad. Sci. USA*, 104, 9346–9351.

Bada, J.L. and Lazcano, A. (2002). Origin of life. Some like it hot, but not the first biomolecules. *Science*, 296, 1982–1983.

Bada, J.L., Chalmers, J., Cleaves H.J. (2016). Is formamide a geochemically plausible prebiotic solvent? *Phys. Chem. Chem. Phys.*, 18, 20085–20090.

Barks, H.L., Buckley, R., Grieves, G.A., Di Mauro, E., Hud, N.V., Orlando, T.M. (2010). Guanine, adenine, and hypoxanthine production in UV-irradiated formamide solutions: Relaxation of the requirements for prebiotic purine nucleobase formation. *ChemBioChem*, 11(9), 1240–1243.

Barone, V., Latouche, C., Skouteris, D., Vazart, F., Balucani, N., Ceccarelli, C., Lefloch, B. (2015). Gas-phase formation of the prebiotic molecule formamide: Insights from new quantum computations. *Monthly Notices of the Royal Astronomical Society*, 453, L31–L35.

Becker, S., Thoma, I., Deutsch, A., Gehrke, T., Mayer, P., Zipse, H., Carell, T. (2016). A high-yielding, strictly regioselective prebiotic purine nucleoside formation pathway. *Science*, 352, 833–836.

Becker, S., Wiedemann, S., Okamura, H., Schneider, C., Iwan, K., Crisp. A., Rossa, M., Amatov. T., Carell, T. (2019). Unified prebiotically plausible synthesis of pyrimidine and purine RNA ribonucleotides. *Science*, 366(6461), 76–82.

Biscans, A. (2018). Exploring the emergence of RNA nucleosides and nucleotides on the early Earth. *Life*, 8, 57.

Bisschop, S.E., Jørgensen, J.K., van Dishoeck, E.F., de Wachter, E.B.M. (2007). Testing grain-surface chemistry in massive hot-core regions. *Astronomy and Astrophysics*, 465, 913.

Biver, N., Bockelée-Morvan, D., Debout, V., Crovisier, J., Boissier, J., Lis, D.C., Dello Russo, N., Moreno, R., Colom, P., Paubert, G., Vervack, R., Weaver, H.A. (2014). Complex organic molecules in comets C/2012 F6 (Lemmon) and C/2013 R1 (Lovejoy): Detection of ethylene glycol and formamide. *Astronomy and Astrophysics*, 566, L5.

Bizzarri, B.M., Botta, L., Pérez-Valverde, M.I., Saladino, R., Di Mauro, E., García-Ruiz, J.M. (2018). A universal geochemical scenario for formamide condensation and prebiotic chemistry. *Chem. Eur. J.*, 24, 8126–8132.

Bizzarri, B.M., Saladino, R., Delfino, I., García-Ruiz, J.M., Di Mauro, E. (2021a). Prebiotic organic chemistry of formamide and the origin of life in planetary conditions: What we know and what is the future. *Int. J. Mol. Sci.*, 22(2), 917.

Bizzarri, B.M., Fanelli, A., Kapralov, M., Krasavin, E., Saladino, R. (2021b). Meteorite-catalyzed intermolecular trans-glycosylation produces nucleosides under proton beam irradiation. *RSC Adv.*, 11, 19258–19264.

Bizzarri, B.M., Fanelli, A., Botta, L., De Angelis, M., Palamara, A.T., Nencioni, L., Saladino, R. (2021c). Aminomalononitrile inspired prebiotic chemistry as a novel multicomponent tool for the synthesis of imidazole and purine derivatives with anti-influenza A virus activity. *RSC Adv.*, 11, 30020–30029.

Bizzarri, B.M., Fanelli, A., Cesarini, S., Saladino, R. (2022a). A three-way regioselective synthesis of amino acid decorated imidazole, purine and pyrimidine derivatives by multicomponent chemistry starting from prebiotic diaminomaleonitrile. *Eur. J. Org. Chem.*, 25, e202200598.

Bizzarri, B.M., Fanelli, A., Ciprini, S., Giorgi, A., De Angelis, M., Fioravanti, R., Nencioni, L., Saladino, R. (2022b). Multicomponent synthesis of diaminopurine and guanine PNA's analogues active against influenza A virus from prebiotic compounds. *ACS Omega*, 7(49), 45253–45264.

Bockelée-Morvan, D., Lis, D.C., Wink, J.E., Despois, D., Crovisier, J., Bachiller, R., Benford, D.J., Biver, N., Colom, P., Davies, J.K. et al. (2000). New molecules found in comet C/1995 O1 (Hale-Bopp). Investigating the link between cometary and interstellar material. *A&A*, 53, 1101.

Botta, L., Saladino, R., Bizzarri, B.M., Cobucci Ponzano, B., Iacono, R., Avino, R., Caliro, S., Carandente, A., Lorenzini, F., Tortora, A., Di Mauro, E., Moracci, M. (2018). Formamide-based prebiotic chemistry in the Phlegrean Fields. *Adv. Space Res.*, 62(8), 2372–2379.

Bruffaerts, J., von Wolff, N., Diskin-Posner, Y., Ben-David, Y., Milstein, D. (2019). Formamides as isocyanate surrogates: A mechanistically driven approach to the development of atom efficient, selective catalytic syntheses of ureas, carbamates, and heterocycles. *J. Am. Chem. Soc.*, 141(41), 16486–16493.

Burton, A.S. and Berger, E.L. (2018). Insights into abiotically-generated amino acid enantiomeric excesses found in meteorites. *Life*, 8(2), 14.

Burton, A.S., Stern, J.C., Elsila, J.E., Glavin, D.P., Dworkin, J.P. (2012). Understanding prebiotic chemistry through the analysis of extraterrestrial amino acids and nucleobases in meteorites. *Chem. Soc. Rev.*, 41, 5459–5472.

Carota, E., Botta, G., Rotelli, L., Di Mauro, E., Saladino, R. (2015). Current advances in prebiotic chemistry under space conditions. *Current Org. Chem.*, 19, 20.

Cataldo, F., Patanè, G., Compagnini, G. (2009). Synthesis of HCN polymer from thermal decomposition of formamide. *J. Macromolecular Sci., Part A Pure and Applied Chemistry*, 46, 11.

Chyba, C.F. (1987). The cometary contribution of the oceans of primitive Earth. *Nature*, 330, 632–635.

Closs, A.C., Fuks, E., Bechtel, M., Trapp, O. (2020). Prebiotically plausible organocatalysts enabling a selective photoredox α-alkylation of aldehydes on the early Earth. *Chemistry*, 26(47), 10702–10706.

Cooper, G.J.T., Surman, A.J., McIver, J., Colón-Santos, S.M., Gromski, P.S., Buchwald, S., Suárez, M.I., Cronin, L. (2017). Miller-Urey spark-discharge experiments in the deuterium world. *Angew. Chem. Int. Ed.*, 56, 8079.

Cossetti, C., Crestini, C., Saladino, R., Di Mauro, E. (2010). Borate minerals and RNA stability. *Polymers*, 2(3), 211–228.

Costanzo, G., Saladino, R., Crestini, C., Ciciriello, F., Di Mauro, E. (2007a). Formamide as the main building block in the origin of nucleic acids. *BMC Evol. Biol.*, 7(Suppl. 2), S1.

Costanzo, G., Saladino, R., Crestini, C., Ciciriello, F., Di Mauro, E. (2007b). Nucleoside phosphorylation by phosphate minerals. *J. Biol. Chem.*, 282, 16729–16735.

Di Mauro, E., Costanzo, G., Negri, R. (1994). One-lane chemical sequencing of PCR amplified DNA: The use of terminal transferase and of the base analogue inosine. *Nucleic Acids Res.*, 22, 3811–3812.

Di Mauro, E., Bizzarri, B.M., Saladino, R. (2021). The role of photochemistry in the prebiotic model of formamide. *Comprehensive Series in Photochemical and Photobiological Sciences*, 20, 107–123.

Emerson, J.W., Green, W.A., Schloerke, B., Crowley, J., Cook, D., Hofmann, H., Wickham, H. (2013). The generalized pairs plot. *J. Comput. Graph. Stat.*, 22, 79–91.

Eschenmoser, A. and Loewenthal, E. (1992). Chemistry of potentially prebiological natural products. *Chem. Soc. Rev.*, 21, 1–16.

Fedoseev, G., Ioppolo, S., Zhao, D., Lamberts, T., Linnartz, H. (2015). Low-temperature surface formation of NH_3 and HNCO: Hydrogenation of nitrogen atoms in CO-rich interstellar ice analogues. *Monthly Notices of the Royal Astronomical Society*, 446, 439–448.

Ferris, J.P., Donner, D.B., Lotz, W. (1972). Chemical evolution IX. Mechanism of the oligomerization of hydrogen cyanide and its possible role in the origins of life. *J. Am. Chem. Soc.*, 94(20), 6968–6974.

Ferus, M., Michalčíková, R., Shestivská, V., Šponer, J., Šponer, J.E., Civiš, S. (2014). High-energy chemistry of formamide: A simpler way for nucleobase formation. *J. Phys. Chem. A.*, 118, 719–736.

Ferus, M., Nesvorný, D., Šponer, J., Kubelík, P., Michalčíková, R., Shestivská, V., Šponer, J.E., Civiš, S. (2015). High-energy chemistry of formamide: A unified mechanism of nucleobase formation. *Proc. Natl. Acad. Sci. USA*, 112, 657–666.

Ferus, M., Pietrucci, F., Saitta, A.M., Knížek, A., Kubelík, P., Ivanek, O., Shestivska, V., Civiš, S. (2017). Formation of nucleobases in a Miller–Urey reducing atmosphere. *Proc. Natl. Acad. Sci. USA*, 114(17), 4306–4311.

Foustoukos, D.I. and Seyfried, W.E. (2004). Hydrocarbons in hydrothermal vent fluids: The role of chromium-bearing catalysts. *Science*, 304, 1002–1005.

Furukawa, Y., Horiuchi, M., Kakegawa, T. (2013). Selective stabilization of ribose by borate. *Orig. Life Evol. Biosph.*, 43(4–5), 353–361.

García-Ruiz, J.M., Hyde, S.T., Carnerup, A.M., Christy, A.G., Van Kranendonk, M.J., Welham, N.J. (2003). Self-assembled silica-carbonate structures and detection of ancient microfossils. *Science*, 302, 1194–1197.

Georgelin, T., Maguy, J., Fournier, F., Laurent, G., Costa-Torro, F., Maurel, M.-C., Lambert, J.-F. (2015). Stabilization of ribofuranose by a mineral surface. *Carbohydr. Res.*, 402, 241–244.

Glaab, F., Rieder, J., García-Ruiz, J.M., Kunz, W., Kellermeier, M. (2016). Diffusion and precipitation processes in iron-based silica gardens. *Phys. Chem. Chem. Phys.*, 18, 24850–24858.

Glaab, F., Rieder, J., Klein, R., Choquesillo-Lazarte, D., Melero-Garcia, E., García-Ruiz, J.M., Kunz, W., Kellermeier, M. (2017). Precipitation and crystallization kinetics in silica gardens. *ChemPhysChem*, 18, 338–345.

Goesmann, F., Rosenbauer, H., Bredehöft, J.H., Cabane, M., Ehrenfreund, P., Gautier, T., Giri, C., Krüger, H., Roy, J., Macdermott, A.J. et al. (2015). Organic compounds on comet 67P/Churyumov-Gerasimenko revealed by COSAC mass spectrometry. *Science*, 349(6247).

Gorb, L., Asensio, A., Tuñón, I., Ruiz-López, M.F. (2005). The mechanism of formamide hydrolysis in water from ab initio calculations and simulations. *Chemistry-A Eur. J.*, 11(22), 6743–6753.

Guo, X., Fu, S., Ying, J., Zhao, Y. (2023). Prebiotic chemistry: A review of nucleoside phosphorylation and polymerization. *Open Biol.*, 13, 220–234.

Hudson, J.S., Eberle, J.F., Vachhani, R.H., Rogers, L.C., Wade, J.H., Ramanarayanan, K., Springsteen, G. (2012). A unified mechanism for abiotic adenine and purine synthesis in formamide. *Angewandte Chemie Int. Ed.*, 124(21), 5224–5227.

James, R.H., Green, D.R.H., Stock, H.J., Alker, B.J., Banerjee, N.R., Cole, C., German, C.R., Huvenne, V.A.I., Powell, A.M., Connelly, D.P. (2014). Composition of hydrothermal fluids and mineralogy of associated chimney material on the East Scotia Ridge back-arc spreading centre. *Geochim. Cosmochim. Acta*, 139, 47–71.

Johnson, A.P., Cleaves, H.J., Dworkin, J.P., Glavin, D.P., Lazcano, A., Bada, J.L. (2008). The Miller volcanic spark discharge experiment. *Science*, 322, 404.

Joussein, E., Petit, S., Delvaux, B. (2007). Behavior of halloysite clay under formamide treatment. *Applied Clay Science*, 35 (1–2), 17–24.

Kahane, C., Ceccarelli, C., Faure, A., Caux, E. (2013). Detection of formamide, the simplest but crucial amide, in a solar-type protostar. *ApJ*, 763, L38.

Kakumoto, T., Saito, K., Imamura, A. (1985). Thermal decomposition of formamide: Shock tube experiments and ab initio calculations. *J. Phys. Chem.*, 89(11), 2286–2291.

Kellermeier, M., Glaab, F., Melero-Garcia, E., García-Ruiz, J.M. (2013). *Methods in Enzymology*. Elsevier, Oxford.

Kim, H.-J., Furukawa, Y., Kakegawa, T., Bita, A., Scorei, R., Benner, S.A. (2016). Evaporite borate-containing mineral ensembles make phosphate available and regiospecifically phosphorylate ribonucleosides: Borate as a multifaceted problem solver in prebiotic chemistry. *Angew. Chem. Int. Ed. Engl.*, 51, 15816–15820.

Kiss, B., Picaud, S., Szőri, M., Jedlovszky, P. (2019). Adsorption of formamide at the surface of amorphous and crystalline ices under interstellar and tropospheric conditions. A grand canonical Monte Carlo simulation study. *J. Physis. Chem. A*, 123(13), 2935–2948.

Lambert, J.B., Lu, G., Singer, S.R., Kolb, V.M. (2004). Silicate complexes of sugars in aqueous solution. *J. Am. Chem. Soc.*, 126, 9611–9625.

Lefloch, B., Ceccarelli, C., Codella, C., Favre, C., Podio, L., Vastel, C., Viti, S., Bachiller, R. (2017). L1157-B1, a factory of complex organic molecules in a solar-type starforming region. *Monthly Notices of the Royal Astronomical Society*, 469, L73–L77.

Lei, M., Ma, L., Hu, L. (2010). A convenient one-pot synthesis of formamide derivatives using thiamine hydrochloride as a novel catalyst. *Tetrahedron Lett.*, 51(32), 4186–4188.

Levy, M. and Miller, S.L. (1998). The stability of the RNA bases: Implications for the origin of life. *Proc. Natl. Acad. Sci. USA*, 95(14), 7933–7938.

Levy, M., Miller, S.L., Brinton, K., Bada J.L. (2000). Prebiotic synthesis of adenine and amino acids under Europa-like conditions. *Icarus*, 145(2), 609–613.

López-Sepulcre, A., Jaber, A., Mendoza, E., Lefloch, B., Ceccarelli, C., Vastel, C., Bachiller, R., Cernicharo, J., Codella, C., Kahane, C. et al. (2015). Shedding light on the formation of the pre-biotic molecule formamide with ASAI. *Monthly Notices of the Royal Astronomical Society*, 449(3), 2438–2458.

López-Sepulcre, A., Balucani, N., Ceccarelli, C., Codella, C., Dulieu, F., Theulé, P. (2019). Interstellar formamide (NH_2CHO), a key prebiotic precursor. *ACS Earth Space Chem.*, 3(10), 2122–2137.

Martin, Z. (2018). The nitrogen heterocycle content of meteorites and their significance for the origin of life. *Life*, 8(3), 28.

Meinhold, G. (2010). Rutile and its applications in earth sciences. *Earth-Sci. Rev.*, 102, 1–28.

Mendoza, E., Lefloch, B., López-Sepulcre, A., Ceccarelli, C., Codella, C., Boechat-Roberty, H.M., Bachiller, R. (2014). Molecules with a peptide link in protostellar shocks: A comprehensive study of L1157. *MNRAS*, 445, 151.

Miller, S.L. (1953). A production of amino acids under possible primitive earth conditions. *Science*, 117, 528–529.

Miyakawa, S., Yamanashi, H., Kobayashi, K., Cleaves, H.J., Miller, S.L. (2002a). Prebiotic synthesis from CO atmospheres: Implications for the origins of life. *Proc. Natl. Acad. Sci. USA*, 99, 14628–14631.

Miyakawa, S., Cleaves, H.J., Miller, S.L (2002b). The cold origin of life: A. Implications based on the hydrolytic stabilities of hydrogen cyanide and formamide. *Orig. Life Evol. Biosph.*, 32, 195–208.

Negri, R., Costanzo, G., Saladino, R., Di Mauro, E. (1996). One-step, one-lane chemical DNA sequencing by *N*-methylformamide in the presence of metal ions. *Biotechniques*, 21, 910–917.

Nelson, D.L. and Cox, M.M. (2000). *Lehninger Principles of Biochemistry*, 3rd edition. Worth Publishers, New York.

Neubeck, A., Duc, N.T., Bastviken, D., Crill, P., Holm, N.G. (2011). Formation of H2 and CH4 by weathering of olivine at temperatures between 30 and 70°C. *Geochem. Trans.*, 12, 6.

Niether, D., Afanasenkau, D., Dhont, J.K.G., Wiegand, S. (2016). Accumulation of formamide in hydrothermal pores to form prebiotic nucleobases. *Proc. Natl. Acad. Sci. USA.*, 113, 4272–4277.

Orgel, L.E. (2004). Prebiotic adenine revisited: Eutectics and photochemistry. *Orig. Life Evol. Biosph.*, 34, 361–369.

Orò, J. (1963). Synthesis of organic compounds by electric discharges. *Nature*, 197, 862–867.

Parker, E.T., Cleaves, H.J., Dworkin, J.P., Glavin, D.P., Callahan, M., Aubrey, A., Lazcano, A., Bada, J.L. (2011). Primordial synthesis of amines and amino acids in a 1958 Miller H_2S-rich spark discharge experiment. *Proc. Natl. Acad. Sci. USA*, 108, 5526–5531.

Parnell, J., Baron, M., Lindgren, P. (2006). Potential for irradiation of methane to form complex organic molecules in impact craters: Implications for Mars, Titan and Europa. *J. Geochem. Expl.*, 89(1–3), 322–325.

Pastero, L., Aquilano, D., Moret, M. (2012). Selective adsorption/absorption of formamide in NaCl crystals growing from solution. *Crystal Growth & Design*, 12(5), 2306–2314.

Pfaffe, M. and Mahrwald, R. (2012). Direct glycosylation of unprotected and unactivated carbohydrates under mild conditions. *Org. Lett.*, 14, 792–795.

Pino, S., Sponer, J.E., Costanzo, G., Saladino, R., Mauro, E.D. (2015). From formamide to RNA, the path is tenuous but continuous. *Life*, 5, 372–384.

Powner, M.W. and Sutherland, J.D. (2010). Phosphate-mediated interconversion of ribo- and arabino- configured prebiotic nucleotide intermediates. *Angew. Chem. Int. Ed.*, 49, 4641–4643.

Powner, W.M., Gerland, B., Sutherland, J.D. (2009). Synthesis of activated pyrimidine ribonucleotides in prebiotically plausible conditions. *Nature*, 459, 239–242.

Price, N.E., Johnson, K.M., Wang, J., Fekry, M.I., Wang, Y., Gates, K.S. (2014). Interstrand DNA–DNA cross-link formation between adenine residues and abasic sites in duplex DNA. *J. Am. Chem. Soc.*, 136, 3483–3490.

Ricardo, A., Carrigan, M.A., Olcott, A.N., Benner, S.N. (2004). Borate minerals stabilize ribose. *Science*, 303, 196.

Rimola, A., Tosoni, S., Sodupe, M., Ugliengo P. (2005). Peptide bond formation activated by the interplay of Lewis and Bronsted catalysts. *Chem. Phys. Lett.*, 408(4–6), 295–301.

Rimola, A., Costa, D., Sodupe, M., Lambert, J.F., Ugliengo, P. (2013). Silica surface features and their role in the adsorption of biomolecules: Computational modeling and experiments. *Chem. Rev.*, 113, 4216–4313.

Rotelli, L., Trigo-Rodríguez, J., Moyano-Cambero, C., Carota, E., Botta, L., Di Mauro, E., Saladino, R. (2016). The key role of meteorites in the formation of relevant prebiotic molecules in a formamide/water environment. *Sci. Rep.*, 6, 38888.

Rubin, R.H., Swenson Jr., G.W., Benson, R.C., Tigelaar, H.L., Flygare, W.H. (1971). Microwave detection of interstellar formamide. *ApJ*, 169, L39.

Ruf, A., D'Hendecourt, L.L.S., Schmitt-Kopplin, P. (2018). Data-driven astrochemistry: One step further within the origin of life puzzle. *Life*, 8(2), 18.

Russell, M.J., Barge, L.M., Bhartia, R., Bocanegra, D., Bracher, P.J., Branscomb, E., Kidd, R., McGlynn, S., Meier, D.H., Nitschke, W. et al. (2014). The drive to life on wet and icy worlds. *Astrobiology*, 14, 308–343.

Saitta, A.M. and Saija, F. (2014). Miller experiments in atomistic computer simulations. *Proc. Natl. Acad. Sci. USA*, 111, 13768–13773.

Saladino, R., Mincione, E., Crestini, C., Negri, R., Di Mauro, E., Costanzo, G. (1996). Mechanism of degradation of purine nucleosides by formamide. Implications for chemical DNA sequencing procedures. *J. Am. Chem. Soc.*, 118, 5615–5619.

Saladino, R., Crestini, C., Mincione, E., Costanzo, G., Di Mauro, E., Negri, R. (1997). Mechanism of degradation of 2'-deoxycytidine by formamide: Implications for chemical DNA sequencing procedures. *Bioorg. Med. Chem.*, 5(11), 2041–2048.

Saladino, R., Crestini, C., Costanzo, G., Negri, R., Di Mauro, E. (2001). A possible prebiotic synthesis of purine, adenine, cytosine, and 4(3H)-pyrimidinone from formamide: Implications for the origin of life. *Bioorg. Med. Chem.*, 9, 1249–1253.

Saladino, R., Ciambecchini, U., Crestini, C., Costanzo, G., Negri, R., Di Mauro, E. (2003). One-pot TiO_2-catalyzed synthesis of nucleic bases and acyclonucleosides from formamide: Implications for the origin of life. *ChemBioChem*, 4, 514–521.

Saladino, R., Crestini, C., Ciambecchini, U., Ciciriello, F., Costanzo, G., Di Mauro, E. (2005a). Synthesis and degradation of nucleobases and nucleic acids by formamide in the presence of montmorillonites. *ChemBioChem*, 5(11), 1558–1566.

Saladino, R., Crestini, C., Neri, V., Brucato, J.R., Colangeli, L., Ciciriello, F., Di Mauro, E., Costanzo, G. (2005b). Synthesis and degradation of nucleic acid components by formamide and cosmic dust analogues. *ChemBioChem*, 6(8), 1368–1374.

Saladino, R., Crestini, C., Busiello, V., Ciciriello, F., Costanzo, G., Di Mauro, E. (2005c). Differential stability of 3'- and 5'-phosphoester bonds in deoxy monomers and oligomers. *J. Biol. Chem.*, 280, 35658–35669.

Saladino, R., Crestini, C., Ciciriello, F., Di Mauro, E., Costanzo, G. (2006a). Differential stability of phosphoester bonds in ribomonomers and ribooligomers. *J. Biol. Chem.*, 281, 5790–5796.

Saladino, R., Crestini, C., Neri, V., Ciciriello, C., Costanzo, G., Di Mauro, E. (2006b). Origin of informational polymers: The concurrent roles of formamide and phosphates. *Chembiochem*, 7(11), 1707–1714.

Saladino, R., Crestini, C., Ciciriello, F., Costanzo, G., Di Mauro, E. (2007). Formamide chemistry and the origin of informational polymers. *Chemistry & Biodiversity*, 4, 694–720.

Saladino, R., Neri, V., Crestini, C., Costanzo, G., Graciotti, M., Di Mauro, E. (2008). Synthesis and degradation of nucleic acid components by formamide and iron sulfur minerals. *J. Am. Chem. Soc.*, 130(46), 15512–15518.

Saladino, R., Crestini, C., Ciciriello, F., Pino, S., Costanzo, G., Di Mauro, E. (2009). From formamide to RNA: The roles of formamide and water in the evolution of chemical information. *Res. Microbiol.*, 160, 441–448.

Saladino, R., Neri, V., Crestini, C., Costanzo, G., Graciotti, M., Di Mauro, E. (2010). The role of the formamide/zirconia system in the synthesis of nucleobases and biogenic carboxylic acid derivatives. *J. Mol. Evol.*, 71, 100–110.

Saladino, R., Brucato, J.R., De Sio, A., Botta, G., Pace, E., Gambicorti, L. (2011a). Photochemical synthesis of citric acid cycle intermediates based on titanium dioxide. *Astrobiology*, 11, 815–824.

Saladino, R., Barontini, M., Cossetti, C., Di Mauro, E., Crestini, C. (2011b). The effects of borate minerals on the synthesis of nucleic acid bases, amino acids and biogenic carboxylic acids from formamide. *Orig. Life Evol. Biosph.*, 41, 317–330.

Saladino, R., Crestini, C., Cossetti, C., Di Mauro, E., Deamer, D. (2011c). Catalytic effects of Murchison material: Prebiotic synthesis and degradation of RNA precursors. *Orig. Life Evol. Biosph.*, 41, 437.

Saladino, R., Botta, G., Pino, S., Costanzo, G., Di Mauro, E. (2012a). Genetics first or metabolism first? The formamide clue. *Chem. Soc. Rev.*, 41, 5526–5565.

Saladino, R., Crestini, C., Pino, S., Costanzo, G., Di Mauro, E. (2012b). Formamide and the origin of life. *Physics of Life Reviews*, 9(1), 84–104.

Saladino, R., Botta, G., Delfino, M., Di Mauro, E. (2013). Meteorites as catalysts for prebiotic chemistry. *Chem. Eur. J.*, 19(50), 16916–16922.

Saladino, R., Carota, E., Botta, G., Di Mauro, E. (2015). Meteorite-catalyzed syntheses of nucleosides and of other prebiotic compounds from formamide under proton irradiation. *Proc. Natl. Acad. Sci. USA*, 112(21), E2746–E2755.

Saladino, R., Botta, G., García-Ruiz, J.M. (2016a). A global scale scenario for prebiotic chemistry: Silica-based self-assembled mineral structures and formamide. *Biochemistry*, 55, 2806–2811.

Saladino, R., Carota, E., Botta, G., Kapralov, M., Timoshenko, G.N., Rozanov, A., Krasavin, E., Di Mauro E. (2016b). First evidence on the role of heavy ion irradiation of meteorites and formamide in the origin of biomolecules. *Orig. Life Evol. Biosph.*, 46, 515–521.

Saladino, R., Bizzarri, B.M., Botta, L., Šponer, J., Šponer, J.E., Georgelin, T., Jaber, M., Rigaud, B., Kapralov, M., Timoshenko, G.N. et al. (2017). Proton irradiation: A key to the challenge of N-glycosidic bond formation in a prebiotic context. *Sci. Rep.*, 7, 14709.

Saladino, R., Šponer, J.E., Šponer, J., Pino, S., Di Mauro, E. (2018a). Chemomimesis and molecular Darwinism in action: From abiotic generation of nucleobases to nucleosides and RNA. *Life*, 8(2), 24.

Saladino, R., Šponer, J.E., Šponer, J., Di Mauro, E. (2018b). Rewarming the primordial soup: Revisitations and rediscoveries in prebiotic chemistry. *ChemBioChem*, 19(1), 22–25.

Saladino, R., Di Mauro, E., García-Ruiz, J.M. (2019). A universal geochemical scenario for formamide condensation and prebiotic chemistry. *Chemistry – A European Journal*, 25(13), 3181–3189.

Sanchez, R.A., Ferris, J.P., Orgel, L.E. (1967). Studies in prebiotic synthesis. II. Synthesis of purine precursors and amino acids from aqueous hydrogen cyanide. *J. Mol. Biol.*, 30, 223–253.

Schoffstall, A.M. (1976). Prebiotic phosphorylation of nucleosides in formamide. *Origins of Life*, 7, 399–412.

Senanayake, S.D. and Idriss, H. (2006). Photocatalysis and the origin of life: Synthesis of nucleoside bases from formamide on $TiO_2(001)$ single surfaces. *Proc. Natl. Acad. Sci. USA*, 103, 1194–1198.

Shapiro, R. (1995). The prebiotic role of adenine: A critical analysis. *Origins Life Evol. Biosphere*, 25, 83–98.

Shapiro, R. (1996). Prebiotic syntheses of the RNA bases: A critical analysis. *Origins Life Evol. Biosphere*, 26, 238–239.

Signorile, M., Salvini, C., Zamirri, L., Bonino, F., Martra, G., Sodupe, M., Ugliengo, P. (2018). Formamide adsorption at the amorphous silica surface: A combined experimental and computational approach. *Life*, 8(42). doi: 10.3390/life8040042.

Slebocka-Tilk, H., Sauriol, F., Monette, M., Brown, R.S. (2002). Aspects of the hydrolysis of formamide: Revisitation of the water reaction and determination of the solvent deuterium kinetic isotope effect in base. *Canadian J. Chem.*, 80, 1343–1350.

Sojo, V., Herschy, B., Whicher, A., Camprubí, E., Lane, N. (2016). The origin of life in alkaline hydrothermal vents. *Astrobiology*, 16(2), 181–197.

Soukup, G.A. and Breaker, R.R. (1999). Relationship between internucleotide linkage geometry and the stability of RNA. *RNA*, 5, 1308–1325.

Šponer, J.E., Mladek, A., Šponer, J., Fuentes-Cabrera, M. (2012). Formamide-based prebiotic synthesis of nucleobases: A kinetically accessible reaction route. *J. Phys. Chem. A*, 116, 720–726.

Šponer, J.E., Szabla, R., Góra, R.W., Saitta, M., Pietrucci, F., Saija, F., Di Mauro, E., Saladino, R., Ferus, M., Civišh, S. et al. (2016a). Prebiotic synthesis of nucleic acids and their building blocks at the atomic level – Merging models and mechanisms from advanced computations and experiments. *Phys. Chem. Chem. Phys.*, 18, 20047–20066.

Šponer, J.E., Šponer, J., Nováková, O., Brabec, V., Šedo, O., Zdráhal, Z., Costanzo, G., Pino, S., Saladino, R., Di Mauro, E. (2016b). Emergence of the first catalytic oligonucleotides in a formamide-based origin scenario. *Chem. Eur. J.*, 22(11), 3572–3586.

Stribling, R. and Miller, S.L. (1987). Energy yields for hydrogen cyanide and formaldehyde syntheses: The HCN and amino acid concentrations in the primitive ocean. *Orig. Life Evol. Biosph.*, 17, 261–273.

Wetmore, S.D., Boyd, R.J., Eriksson, L.A. (1998). Theoretical investigation of adenine radicals generated in irradiated DNA components. *J. Phys. Chem. B*, 102, 10602–10614.

Yamada, H., Hirobe, M., Higashiyama, K., Takahashi, H., Suzuki, K.T. (1978a). Reaction mechanism for purine ring formation as studied by 13C-15N coupling. *Tetrahedron Lett.*, 42, 4039–4044.

Yamada, H., Hirobe, M., Higashiyama, K., Takahashi, H., Suzuki, K.T. (1978b). Detection of carbon-13-nitrogen-15 coupled units in adenine derived from doubly labeled hydrogen cyanide or formamide. *J. Am. Chem. Soc.*, 100, 4617–4618.

Zubay, G. and Mui, T. (2001). Prebiotic synthesis of nucleotides. *Orig. Life Evol. Biosph.*, 31, 87–102.

4

A Praise of Imperfection: Emergence and Evolution of Metabolism

Juli PERETÓ[1,2]
[1] *Institute for Integrative Systems Biology (I2SysBio), CSIC, Universitat de València, Spain*
[2] *Department of Biochemistry and Molecular Biology, Universitat de València, Spain*

Dedicated to the memory of Josep Casadesús (1952–2022) and Dan Salah Tawfik (1955–2021).

4.1. From Darwin to Jacob: perfection does not exist

Nothing is perfect and the illusion of perfection only exists in the minds of humans. For natural theologians, the apparent perfection of living things, with organs very well suited to their functions, was a demonstration of the existence of a designer. In his undergraduate years at the University of Cambridge, Charles R. Darwin was a great admirer of the Reverend William Paley and his arguments for divine design and finality in nature. However, as Darwin matured his ideas on natural and sexual selection, he amassed an enormous number of observations and refined the arguments for a design without a designer. Indeed, as Català-Gorgues (2022) has pointed out, without abandoning finalist arguments altogether, he negotiated with the design argument to provide an alternative mechanism to the creator-driven purpose that Darwin considered to be nonexplanatory. If we carefully read his various works for philosophical subtleties, as Ghiselin (1969) masterfully did, we will find a sophisticated thinking that refutes, sometimes with ironic touches, a providential design. Consider the case of the eye that Paley had proposed as an example of an organ of extreme complexity for which, necessarily, there had to be someone who anticipated the function and purpose, an expert in optical science and engineering, like someone who designs and builds a telescope. Darwin, who not only thought rigorously and creatively, but subjected his ideas to a strict self-criticism, used the example of the eye in the chapter devoted to the serious issues of his theory in *On the Origin of Species*. Although in the first edition of this book (1859), Darwin dropped his idea of optical perfection, derived from the theological illusion (see Chapter 6 "Difficulties on theory", in particular the section "Organs of *extreme perfection*", (author's emphasis), his mind gradually changed until the sixth and definitive edition (published in 1872), in which his doubts on the supposed perfection of the eye were already clearly expressed (see the evolution of this topic over the six editions in Domínguez (2017)). Very explicitly, in the second edition of *The Descent of Man* (1874) he resorts to the opinion of the physicist Hermann von Helmholtz to definitively support the idea of imperfection as the outcome of the evolutionary process (see also Fishman 2010):

> We have, however, no right to expect absolute perfection in a part rendered ornamental through sexual selection, any more than we have in a part modified through natural selection for real use; for instance in that wondrous organ the human eye. And we know what Helmholtz, the highest authority in Europe on the subject, has said about the human eye; that if an optician had sold him an instrument so carelessly made, he would have thought himself fully justified in returning it (Darwin 1874, p. 441).

But the work in which Darwin radically dismantles the design argument is in his monograph on orchid pollination. After Darwin, no one could see a flower as an object for the purpose of delighting our senses. Flowers are the outcome of an elaborate process of coevolution with pollinating insects and, for Darwin, the delicate gears and devices we observe are not the result of any prior plan but of the recycling of old parts with different functions, recruited for a new purpose. This *"flank movement" on the enemy*, as Darwin (1862a) called it, has been highlighted by Ghiselin (1969) as the most forceful and definitive attack on the design argument: if the present function is derived, occasionally, opportunistically, imperfectly from the combination of devices with different former functions, how could there have been purpose originally?

It is worth recalling Darwin's own words on the matter:

> Although an organ may not have been originally formed for some special purpose, if it now serves for this end we are justified in saying that it is specially contrived for it. On the same principle, if a man were to make a machine for some special purpose, but were to use old wheels, springs, and pulleys, only slightly altered, the whole machine, with all its parts, might be said to be specially contrived for that purpose. Thus throughout nature almost every part of each living being has probably served, in a slightly modified condition, for diverse purposes, and has acted in the living machinery of many ancient and distinct specific forms (Darwin 1862b, p. 348).

Probably without having yet read this fragment of Darwin – later perhaps through the reading of Ghiselin's book – François Jacob offered a wonderful intellectual convergence with the British naturalist by putting forward his idea of molecular bricolage or tinkering. What Darwin proposed for orchids, Jacob applied to molecules. The outstanding bricolage metaphor was first published in 1977 in *Science*, in what was a summary of a lecture given months earlier at the University of California, Berkeley. Jacob enriched his idea with more examples and details over the following years, in texts such as *Le jeu des possibles* (1981) or "Molecular tinkering in evolution" (1983), as his contribution to a volume commemorating the centenary of Darwin's death. In his 1981 text, he already points out Darwin's priority in offering this sloppy and opportunistic view of evolution that he recognized in the evolution of many proteins, such as globins or collagen, in genes fragmented by introns or in the existence of mobile elements in the genome.

Furthermore, Jacob proposed that, in prebiotic times, very simple building blocks emerged by chance and the assembly of short peptides generated the first functional proteins. In fact, this idea is reminiscent of the pioneering proposal of Margaret Oakley Dayhoff on the origin of ferredoxin by tandem duplications of a simpler ancestral peptide emerged from the condensation of abiotic amino acids (Eck and Dayhoff 1966). According to Jacob, the creative phase of biochemistry must have occurred very early, which gave rise to the extraordinary molecular unity observed in living beings. In later evolutionary phases, it would be the ecological circumstances that would shape the molecular parts, recycling old devices for new functions, altering the specialization and organization of those parts. Each level of organization has its own constraints that mark the limit of what is possible. It is the combination of physicochemical determinism and history, in varying proportions, that delimits evolutionary possibilities. The weight of history, however, will increase as the complexity of the system increases: living things are historical creatures shaped by this cocktail of bricolage and opportunism.

For the historian Michel Morange (2002), the metaphor of bricolage is the richest of all those proposed by Jacob throughout his long work, an image that still retains its full value today in the context of our current knowledge of molecular and biochemical evolution. On the other hand, Víctor de Lorenzo (2018) has analyzed the nuances of the application of this metaphor to the field of synthetic biology. My purpose in this chapter is to explore the actual state of the study of the origin and evolution of metabolism – including the artificial one – in light of these notions of Darwin and Jacob.

4.2. Protometabolic networks

A cloud hides the connection between prebiotic chemistry and present-day metabolism, in a powerful image suggested by Antonio Lazcano and Stanley L. Miller in 1999. The cloud represents the difficulty of understanding how the transition from protometabolic networks – before the emergence of enzymatic catalysts – to more ancient metabolic pathways occurred. Looking in detail for possible parallels in specific mechanisms, there are some cases in which the enzymatic steps of metabolism resemble abiotic transformations. In this context, we speak of chemomimetic biosynthesis. But in fact, the repertoire of possible chemomimetic reactions in metabolism identified so far is disappointingly poor (reviewed in Peretó 2012).

This lack of strict parallelism does not preclude a principle of congruence between a (nonenzymatic) protometabolism and a modern (enzymatic) metabolism, as proposed by Christian de Duve (for a summary of his ideas, see de Duve 2005). He assumed that there was a chemical determinism in the origin of metabolism. For de Duve, the term protometabolism is synonymous with prebiotic chemistry, that is, the early chemical processes that preceded and gave rise to existing metabolism, and the term has also recently been adopted by some researchers in systems chemistry. Protometabolism would be the ideal chemical space for the emergence of useful catalysts in an RNA world. Replacing protometabolism with modern metabolism would require the emergence of catalysts by a selection process based on their usefulness under certain conditions. For example, in an RNA world, only ribozymes that match existing substrates would be selected to transform them into suitable products, since an evolving catalyst would be functionally meaningful as such only if a suitable substrate exists and is ready to be transformed into a product that benefits the overall system. The initial set of substrates would be the product of more primitive catalysts: mineral surfaces, metal or inorganic ions, small organic molecules, such as amino acids or short polymers, such as the noncoding peptides that de Duve called multimers, lipid vesicles, etc.

The emergence of ribozymes, as subjects of natural selection, added a new dimension to pure chemical processes, as adaptation allowed for the optimization of catalytic traits. Thus, the first enzymatic catalysts under natural selection (whether ribozymes in an RNA world or, later, genetically encoded enzymatic proteins) would play a key role in the transition from a "dirty" (de Duve's "gemisch") or unfocused protometabolism to a sharper metabolic map. But how "dirty" was the abiotic chemical landscape? Results from prebiotic systems chemistry indicate that chemical determinism would lead to a more restricted abiotic repertoire than previously imagined.

Thus, Lazcano and Miller's (1999) cloud began to dissipate in recent years as there are some pieces of the puzzle that fit de Duve's definition of protometabolism. Chemical systems have been explored that result in the formation of compounds that are the same or very similar to those involved in current metabolism. For instance, John Sutherland proposed the cyanosulfidic protometabolism (Patel et al. 2015) as a network of nonenzymatic reactions leading to the synthesis of precursors of protein amino acids, nucleotides and lipids, starting from very simple molecules (HCN, H_2S, phosphate) and ultraviolet light. Sutherland's protometabolism is a good example of what is now known as systems chemistry, the exploration of heterogeneous systems in which the participant chemicals can also be catalysts for some of the observed processes.

The groups of Joseph Moran and Markus Ralser have studied nonenzymatic transformations, catalyzed by iron and other transition metals, around the citric acid cycle (CAC), a catalytic cycle of central metabolism that provides energy by oxidation of two-carbon molecules to carbon dioxide and supplies essential intermediates as a starting point for biosynthesis, and perhaps one of the oldest parts of metabolism. Also, the formamide chemistry of Raffaele Saladino and Ernesto Di Mauro, catalyzed by minerals and meteoritic materials, reconstructs much of the CAC. What is interesting is that CAC involves a number of six-, five-, and four-carbon intermediates that in turn may be precursors of multiple essential cellular components, such as amino acids, and from these, matter could circulate toward the synthesis of nitrogenous bases. For his part, Matt Powner has described a series of abiotic reactions that mimic the second part of glycolysis, an almost universal route of transformation of sugars into three- or two-carbon intermediates, ready for incorporation into more central routes such as CAC (for specific references on the above works see Peretó 2019).

Of all the recent protometabolic explorations, perhaps the most promising is the one initiated by the groups of Ram Krishnamurthy and Greg Springsteen, which moves through the chemical neighborhood of CAC and has the particularity of not requiring metal catalysts. Starting from the observation that transition metals promote not only the formation of molecules but also their rapid destruction, and following a systemic strategy under aqueous conditions and mild temperature, Krishnamurthy and Springsteen have described several cycles, possible precursors of metabolism with chemical parallels to modern biochemical strategies, but using simple feedstock reagents, such as HCN or glyoxylate. In a series of papers, these researchers have shown possible pathways to amino acids and nitrogenous bases through nonenzymatic reactions, which are plausible in a prebiotic world. These observations offer simpler alternatives to current metabolic mechanisms and represent a model of reactions that could be operative before the appearance of complicated protein enzymes and their sophisticated cofactors and would pave the way to more complex metabolic cycles (Springsteen et al. 2018; Stubbs et al. 2020; Pulletikurti et al. 2022).

There is a remarkable epistemological aspect to Krishnamurthy and Springsteen's approach. Research in prebiotic evolution has been strongly influenced by hypotheses rooted in known biology: what would be the origin of the amino acids present in extant proteins, of RNA precursors, of membrane lipids? This extrapolation of biochemistry into the past, in search of its prebiotic precursors, may have unconsciously limited research. Because this focus has diverted attention from the role that other relevant molecules that are not present in the evolved biochemistry could have played in the primitive stages. Protometabolic reactions that explore the chemical neighborhood of biological pathways such as CAC, noncanonical nitrogenous bases (such as orotic acid), the depsipeptide precursors of polypeptides, or hybrid polymers of RNA and DNA, all could be telling us about prebiotic processes before the choices that led to universal biochemistry, possible essential elements, necessary during the process, but absent in the final product (Krishnamurthy 2020).

The empirical approaches to protometabolism are also great news after decades of skepticism about the possibility of observing reaction networks that might represent primitive, geochemically reasonable chemistries, prior to the adoption of enzymatic catalysis (Orgel 2000; Schwartz 2007; Luisi 2014). The current experiments clearly represent a triumph of systems chemistry. The subset of products of protometabolism explored so far is congruent with the diversity of the simplest biomolecules and includes other related molecules absent in universal metabolism. Most of the nonenzymatic transformations between metabolites or their analogues reproduce the general topology of central metabolism, with mechanistic strategies comparable to those of present-day metabolism, although far removed from the intricate catalysis of enzymes and their cofactors.

Nevertheless, true abiotic autocatalytic cycles have yet to be discovered, that is, cyclic sequences of reactions that generate more product than is consumed, and that allow the system to grow, as occurs in real metabolism (Xavier et al. 2020; Xavier and Kauffman 2022). As interesting as the research on the prebiotic plausibility of some metabolic intermediates and their nonenzymatic transformations may be, we urgently need examples of real sustainable nonenzymatic catalytic cycles as a model of the primitive stages of the emergence of metabolic cycles. In any case, today more than ever, we have solid empirical reasons to support de Duve's congruence view of a protometabolism prefiguring metabolism.

One of the central questions that remains to be solved is the origin of biocatalysts in the transition from a nonenzymatic network to one with more or less complex macromolecular catalysts under natural selection, either ribozymes or protein enzymes. Systems chemistry is based on the catalytic activities of organometallic compounds, mineral surfaces or certain components of the system itself. For example, phosphate is both substrate and catalyst in Sutherland's cyanosulfidic protometabolism (Patel et al. 2015). It has been proposed that in an RNA world not only ribozymes but also peptides and cofactor precursor molecules (Fried et al. 2022) or lipid vesicles (Hanczyc and Monnard 2017) could be essential for catalytic diversity. Consistent with the scenario proposed by de Duve in which protometabolism is the chemical landscape on which the most primitive catalysts develop, it has been suggested that there was a coevolution of substrates and catalysts, starting with pure ligand recognition and binding processes. A recruitment of abiotic peptides for their ability to bind particular substrates (generalist or ambiguous catalysts) would also combine with a certain catalytic promiscuity (see Noor et al. 2022 and references therein). For example, the widespread presence of phosphate as a component of substrates and coenzymes, but without direct involvement in catalysis, could suggest the prevalence of a selection mechanism based on the recognition of this chemical group. This idea is supported by the fact that some enzyme superfamilies that can be traced back to times before the last universal ancestor share phosphate-binding structural domains, as is the case of the ubiquitous Rossmann fold (Laurino et al. 2016).

The idea that complex enzymes were built from simpler fragments is not new and has already been used by Jacob (1981) in his discussion on molecular tinkering and biochemical evolution. As mentioned above, this notion was popularized by Dayhoff's pioneering work with ferredoxin (Eck and Dayhoff 1966). She noticed the existence of short symmetrical repeated segments in the structure of this admittedly ancient protein. Current data, with numerous ferredoxin sequences and much more powerful structural and phylogenetic analyses than this bioinformatics pioneer could have imagined, have given further credence to this hypothesis (Romero Romero et al. 2016). This internal symmetry that has been observed in numerous enzyme families and that has been considered a trace of their origin from short, assembled, multiplied and diversified peptides has also been discovered in the innermost heart of the ribosome. The tireless work of Ada Yonath has shown that an ancestral duplication of an oligoribonucleotide may have marked the origin of the peptidyl transferase ribozyme responsible for the formation of the peptide bond during the synthesis of proteins (Davidovich et al. 2010; Bose et al. 2022). An innovation of the RNA world that decreed its own demise at the hands of the evolution of protein enzymes.

4.3. Enzyme promiscuity and metabolic innovation

Our view of how enzymes recognize their substrates has changed over time. Early observations of enzyme specificity led Emil Fischer to propose that the relationship between the enzyme and its substrate was like that of a key to its lock (Fischer 1894). As knowledge of enzyme function, its protein nature and three-dimensional structure grew, the original model was accommodated until the current ideas incorporated a more plastic and dynamic notion of substrate recognition. Native enzymes, with a recognized primary catalytic function, are now thought to be in equilibrium, in the absence of substrate, with three-dimensional conformations that are often in short supply. These alternative structures are a sign of the plastic character of the protein structure (as opposed to the rigidity of a lock) and may have the ability to recognize ligands with alternative structures. These minor, non-native activities are called *promiscuous*, a term that, although it may seem unfortunate, has become the accepted one. Promiscuous activities thus represent minor functionalities of enzymes that are not under selective pressure. However, numerous experimental and theoretical studies have pointed out that this promiscuity can also be an opportunity for functional innovation (Khersonsky and Tawfik 2010).

I prefer to explain enzymatic promiscuity to my students with a personal anecdote. A few years ago, I attended a concert of early music by Jordi Savall in the church of the monastery of Poblet in Catalunya. The concert took place in darkness, except for the illumination of the performers. At one point, I was curious to know the title of the piece they were playing and I used my cell phone to illuminate the program. At a time when cell phones did not yet have a flashlight function, the light from the screen was enough to read the text. Obviously, I had not selected my cell phone for the light emitted by its screen, but in very specific circumstances, that light was useful. Similarly, under certain environmental conditions, promiscuous enzymatic activities, not subject to selection, can become adaptive and open the way to functional innovation.

Protein promiscuity – observed, for example, in enzymes, transporters or receptors – has a key role in biochemical evolution. This idea belongs to a new view of protein evolution in which mechanistic and evolutionary explanations enrich each other, as suggested by Morange (2017). The unfortunate disconnection between biochemistry and evolutionary biology (Cornish-Bowden et al. 2014) has been overcome by a functional synthesis combining powerful methodological approaches such as experimental evolution and the synthesis of ancestral proteins by means of the phylogenetic reconstruction of their most probable sequences, also known as protein resurrection (Dean and Thornton 2007).

In a metabolic context, it is accepted that a general mechanism for the evolutionary diversification of enzyme functions, starting from a primitive metabolic network with few catalysts of low specificity, has been gene duplication and divergence. The patchwork model of metabolic evolution proposes that primitive, low-specific or generalist enzymes would be progressively replaced by more specific, specialist enzymes (Waley 1969; Yčas 1974; Jensen 1976; Kacser and Beevy 1984). Most likely, primitive generalist enzymes also coexisted with nonenzymatic transformations, as proposed by the semi-enzymatic model of Lazcano and Miller (1999), a more formal version of the previous patchwork assembly model (Noda-Garcia et al. 2018; Becerra 2021).

However, although it is more or less explicitly admitted in the above models that the evolutionary process will lead to the best performance in terms of the recognition of substrates (or in catalytic capacity), the reality shows that this result, that of highly kinetically capable enzymes, has not always occurred or, rather, has only rarely occurred. The prevalent reductionist view of biochemistry led Jeremy R. Knowles in the 1970s to propose the concept of the "perfect enzyme" for biocatalysts that transformed substrate into product as soon as the substrate reached the catalytic site. In other words, perfect enzymes were only limited by the rate of encounter between the enzyme and its substrate, which would depend on physicochemical conditions. Knowles, together with his colleague W. John Albery, provided empirical data to show that at least in vitro the glycolytic enzyme triose phosphate isomerase was a 'perfect enzyme' (Albery and Knowles 1976). However, several authors pointed out that catalytic perfection may not be the goal of natural selection (see, for example, Benner 1989) and that enzymes should be "good enough" to perform their functions under certain environmental conditions. "Good enough" means here "sufficient" for a function that may lead to better chances of survival and reproduction. In fact, a survey of kinetic parameter values of known enzymes shows that the supposed catalytic perfection is rather an exception, as the vast majority of biocatalysts show values of substrate affinity or catalytic capacity that are far from perfect (Bar-Even et al. 2011). Even worse, under physiological conditions, only one in 10^4 encounters of substrate and enzyme are productive (Bar-Even et al. 2015). This may be due, at least in part, to the proven fact that enzymes (and also their substrates) are conformationally diverse and exist as dynamic populations of three-dimensional structures.

Thus, real enzymes exhibit a wide range of kinetic characteristics that are the result of complex trade-offs between the rate of catalysis and the precision with which they act (Tawfik 2014). Clearly, we find examples where precision has to be as high as possible, at the cost of paying a considerable energetic price. For example, enzymes responsible for the covalent binding of an amino acid to its corresponding tRNA have to be very accurate in their recognition, because in the process of protein synthesis, a mistake in this assignment can no longer be detected and will lead to a substitution of an amino acid when the error occurs. Some of these enzymes are equipped with error-checking and proofreading mechanisms that require energy expenditure in the form of ATP. At the other extreme, we find enzymes that not only do not require such levels of fidelity but display a lack of precision as an advantage. The logic of cellular economy has led to the selection of less precise catalysts, as in the case of enzymes responsible for the synthesis of families of compounds such as membrane phospholipids (for a study of the distribution of enzyme specificity in *Escherichia coli*, see Nam et al. (2012)).

At present, a variety of empirical evidence – mostly coming from the outstanding research of Dan S. Tawfik and his co-workers – indicate that divergence and specialization of homologous enzymes is a major mode of generation of new functions (James and Twafik 2003; Khersonsky and Tawfik 2010; Noda-Garcia et al. 2018). An enzyme may have one major native function and one or more promiscuous functions. Sometimes promiscuous activities may reflect ancient functions that have not been completely erased in the final evolutionary stage, as has been shown by protein resurrection (Dean and Thornton 2007; Khersonsky and Tawfik 2010). After gene duplication, the accumulation of point mutations can alter conformational equilibria, and what was once a minority form may now be the major one, transitioning from one function to another. This allows the adoption of a new function in the face of changing environmental conditions and under different selective pressures.

Perhaps one of the planetary changes with the greatest impact on metabolism was the oxygenation of the Earth. This environmental revolution offers good examples of metabolic innovation. Molecular oxygen of photosynthetic origin progressively accumulated in the oceans until it began its passage into the atmosphere, concluding with an initial increase in oxygen levels to 10% of present levels. This occurred about 2.4 billion years ago and is a time that has been geologically characterized as the great oxidation event (GOE). Oxygen-using enzymes appeared by tinkering from oxygen-independent predecessor enzymes (mostly oxidoreductases that switched substrates requiring only subtle changes in the catalytic site), but there was also a major element of innovation from enzymes with reactions other than redox catalysis or even non-catalytic proteins (Jabłońska and Tawfik 2022). The overall result was a great deal of enzymatic and metabolic innovation that in many cases took opportunistic advantage of the availability of a substrate that gave greater thermodynamic impetus to the reaction. In other cases, the result was the exploration of a completely new chemical space, such as the synthesis of steroids or the catabolic routes of use of aromatic compounds. Phylogenetic studies indicate that many oxygen-dependent enzyme families predate the GOE by more than 500 Ma, probably because they were ancestrally involved in detoxification processes of oxygen of photochemical or intracellular origin (Jabłońska and Tawfik 2021). Even the last common ancestor was endowed perhaps with oxygen-detoxifying enzymes (Valenti et al. 2022).

Empirical and theoretical studies have shown that an enzyme cannot be optimal in all its possible functions and that there is a trade-off between them, which can be eloquently expressed in the phrase "Jack of all trades, master of none". Most importantly, however, it is possible to explore these functions and to make one prevail over the others. This is the strategy followed by the experimental evolution of enzymes, which exploits this plasticity of proteins and allows the selection of one function over the rest. It is even possible to explore the world of the chemically possible, beyond what natural selection has achieved, by making artificially evolved enzymes catalyze new-to-nature reactions, as the work of Frances H. Arnold and her collaborators has shown (Arnold 2022).

The existence of promiscuous enzyme activities results in an "underground messy metabolism". D'Ari and Casadesús (1998) discussed the role of promiscuous activities on cellular metabolites and the existence of an "underground metabolism" that under certain circumstances could confer an evolutionary advantage. Tawfik referred to the "messy" nature of biology in that there is an inevitable presence of noise, stochastic variations or promiscuous reactions arising from the inherent flexibility of enzymes (Tawfik 2010), a messiness that can be projected from molecules to organisms to ecosystems. The messiness of the underground metabolism – it is inevitable to see the parallels with pleiotropy at the genomic level – can be the raw material for physiological and evolutionary adaptations. In the following paragraphs, we will look at a few examples.

Gene overexpression in *E. coli* allowed Kim et al. (2010) to discover up to three alternative routes for the biosynthesis of the coenzyme pyridoxal-5'-phosphate in a strain with the native route obstructed by a mutation in an essential step. What was achieved in this case was, by increasing the expression of certain genes, to bring up underground reactions whose serendipitous combination rescued the mutation by causing metabolic fluxes to flow through alternative pathways. Even under physiological conditions, promiscuous reactions can also short-circuit mutations that prevent normal flow through an essential pathway, as shown by Perchat et al. (2022) in an *E. coli* strain with a mutation that blocks the synthesis of β-alanine, a necessary ingredient of coenzyme A biosynthesis. What could be demonstrated in this case is that enzymes with known native functions in central metabolism use alternative substrates that establish a lateral connection from polyamine synthesis to β-alanine production by an alternative route, thus allowing cell viability. The existence of alternative pathways built on promiscuous activities must be considered to understand the non-essentiality of some genes coding for enzymes of central pathways (Kim and Copley 2007). The question of how promiscuous activities can establish connections between different parts of the metabolic network and eventually enable adaptation to environmental changes has been explored at a systemic level by Notebaart et al. (2014). Using information on promiscuous reactions in *E. coli* enzymes available in public databases, a metabolic model was constructed that predicted that 45% of promiscuous reactions connected different metabolic modules and could generate new metabolic pathways. An estimated 20% of these reactions allowed the cell to adapt to alternative carbon sources.

More generally, using computational approaches to sample random metabolic networks to sustain life from a given carbon source, Barve and Wagner (2013) showed that metabolism entails multiple nonadaptive traits that allow functional innovation: the selection of networks that allow growth on glucose as the sole carbon source also allows growth on more than 40 alternative substrates. This is what in evolutionary biology is called exaptation, also known as preadaptation (although some people do not like this word because it is an old and useless term): adopting new functions using traits not subject to selection, a further example of the generation of new functions from old, nonfunctional traits or traits formerly with different functions, as suggested by Darwin and Jacob. In a nutshell, promiscuous activities are opportunities stored in the attic of evolution.

These and many other published data show that promiscuous reactions are ubiquitous and represent a functional potential that allow metabolic adaptations and innovations. However, promiscuous reactions also show a dark side when their products are molecules that are toxic to the cell itself. In this case, what has been observed is that cellular metabolism has been provided with repair mechanisms. We are very familiar with the need for repair at the genomic scale, with a variety of error correction and damage repair mechanisms that keep genetic information within mutation tolerance limits adjusted by natural selection. Evolution has also provided cells with metabolic damage repair mechanisms (Hanson et al. 2016). A well-known example is a widely distributed phosphatase that assists in the destruction of the by-products of two enzymes of the glycolytic pathway, whose toxic effects, if accumulated, would be lethal to the cell as inhibitors of key enzymes of central metabolism (Beaudoin and Hanson 2016; Collard et al. 2016). Considering that even repair enzymes exhibit side reactions, this is a good example of how, at times, evolution has solved the imperfections of its own products with imperfect solutions that could hardly be part of a well-designed plan.

4.4. Promiscuity, moonlighting and the essence of life

The question of the nature of life was put aside during the development of molecular biology in the second half of the 20th century (Morange 2010). However, in parallel and disconnected from each other, several scientists maintained an interest in answering the question "what is life?". Research on autopoiesis by Humberto Maturana and Francisco Varela, Manfred Eigen's hypercycle, Tibor Gánti's chemoton, Stuart Kauffman's autocatalytic sets or Robert Rosen's (M,R) systems are a sample of these efforts, a mosaic of explanations, sometimes complementary, on the nature of biological phenomena (for a complete review on these and more models, see Cornish-Bowden and Cárdenas (2020)).

Living beings are systems open to the flow of matter and energy but closed to efficient causes. Unlike machines created by humans, living organisms produce internally the causes of their functioning, their generation and eventual repair or replacement of their component parts. Enzymes are not taken from outside by the cells, but are products of the cells' own metabolism following instructions encoded in the genomes. This metabolic closure is possible because the fundamental problem of *infinite regression* was solved from the beginning, as Rosen pointed out (summarized by Rosen 1991). A biological system endowed with catalysts works as long as the way in which these catalysts, which are material entities with limited lifetimes, are synthesized and their replacement is guaranteed to compensate, for instance, dilution by growth or deterioration and decay. The synthesis of these catalysts will be performed by other catalysts, in turn, with a given half-life. If we do not solve this succession of catalysts that catalyze the synthesis and substitution of other catalysts, we will be imprisoned in an infinite regression.

Rosen approached this issue in a very abstract way, without concessions nor concrete references to biological components. Fortunately, Athel Cornish-Bowden, María Luz Cárdenas and their collaborators have deciphered Rosen's proposals and have provided us with a comprehensible picture of the resolution of infinite regression. According to these authors, the key lies in moonlighting (Cornish-Bowden et al. 2007). Catalysts do not have to be strict functional specialists, some of them have to be endowed with more than one function in the system, so that with a discrete number of catalytic elements, all of the tasks necessary for the existence of the system can be solved.

This notion endows moonlighting (the coexistence of more than one catalytic or non-catalytic function in the same protein) and functional promiscuity with a fundamental role in explaining essential aspects of living things. That enzymes can have more than one recognized function beyond their catalytic activity is now well known (Jeffery 2009). Classical examples, also cited by Jacob (2001), include the recruitment of metabolic enzymes to form part of the lens of the eye, such as the duck δ-crystallin which is also an enzyme of the urea cycle (Piatigorsky et al. 1988). The resolution of the infinite regression by moonlighting catalysts shows us that multitasking is not just an oddity but a necessity harbored in biological systems since their origin and essential for their very existence. Moonlighting and promiscuity can help to answer key questions about the origin and evolution of metabolism. But what may be a dream for evolutionary biologists can become a nightmare for a harder, engineering vision of synthetic biology.

Living beings are not machines. Far from the Cartesian mechanistic ideal, there are several fundamental reasons that differentiate cells from machines, as both scientists and philosophers have pointed out (Nicholson 2014, 2019; Porcar and Peretó 2016). In the context of this chapter, an essential difference between living things and machines is precisely the fact that the efficient causes of living things are internal. As we have mentioned, this implies that there are cellular components that manifest a functional degeneration that resolves the paradox of infinite regression. Far from being composed of rigid, totally predictable and deterministic elements, cellular components exhibit functional promiscuity, structural plasticity and a fluidity loaded with stochasticity and noise. All of these ingredients are difficult to fit into an engineering ideal that manages standardized parts that function independently of context and lack the capacity, beyond ageing and wear, to dynamically evolve over time. As one of the considered founders of contemporary synthetic biology, Drew Endy, sarcastically pointed out, engineers abhor the emergent properties of a complex living system (Edge 2008). This harder engineering vision of synthetic biology is caught between the desirable and the unattainable.

Life, with all its amazing imperfections, is the result of molecular bricolage, innovating by recycling old parts and giving them new functional meaning. This view serves as inspiration for another synthetic biology strategy that uses evolution as an ally (de Lorenzo 2018; Porcar 2019). Numerous current protein and metabolic engineering projects are showing the way to address the intrinsically disordered nature of biology and discover possible solutions not found by natural selection, going into what some have called *transmetabolism* (Peretó 2021). The strategy consists of finding complex solutions, which depend on many parameters, by interrogating the evolution of the system in the test tube.

In the words of the late Dan S. Tawfik (2010):

> Behind the façade of perfection and optimality lies messy biology that originates from evolution and provides the basis for the evolution of all living forms.

This is also a profound truth for synthetic metabolism: imperfection pervades the entire evolutionary process, including the artificial one.

4.5. Acknowledgments

I would like to thank my colleagues Jaume Bertranpetit, Víctor de Lorenzo, Carlos García-Ferris and Manuel Porcar for kindly reading earlier versions of this manuscript (although none of them can be held responsible for any errors). I am also very grateful for having been able to present some of these issues at LSC-Retreat 4.0, Canfranc Underground Laboratory, and for the fruitful debate that arose among the participants, thanks to the invitation of the organizers Tomás Lázaro and Daria Stepanova, and the director of the LSC, Carlos Peña-Garay.

4.6. References

Albery, W.J. and Knowles, J.R. (1976). Evolution of enzyme function and the development of catalytic efficiency. *Biochemistry*, 15, 5631–5640.

Arnold, F.H. (2022). Innovation by evolution: Bringing new chemistry to life. *Biophys. J.*, 121, 177a.

Bar-Even, A., Noor, E., Savir, Y., Liebermeister, W., Davidi, D., Tawfik, D.S., Milo, R. (2011). The moderately efficient enzyme: Evolutionary and physicochemical trends shaping enzyme parameters. *Biochemistry*, 50, 4402–4410.

Bar-Even, A., Milo, R., Noor, E., Tawfik, D.S. (2015). The moderately efficient enzyme: Futile encounters and enzyme floppiness. *Biochemistry*, 54, 4969–4977.

Barve, A. and Wagner, A. (2013). A latent capacity for evolutionary innovation through exaptation in metabolic systems. *Nature*, 500, 203–206.

Beaudoin, G.A. and Hanson, A.D. (2016). A guardian angel phosphatase for mainline carbon metabolism. *Trends Biochem. Sci.*, 41, 893–894.

Becerra, A. (2021). The semi-enzymatic origin of metabolic pathways: Inferring a very early stage of the evolution of life. *J. Mol. Evol.*, 89, 183–188.

Benner, S.A. (1989). Enzyme kinetics and molecular evolution. *Chem. Rev.*, 89, 789–806.

Bose, T., Fridkin, G., Davidovich, C., Krupkin, M., Dinger, N., Falkovich, A.H., Peleg, Y., Agmon, I., Bashan, A., Yonath, A. (2022). Origin of life: Protoribosome forms peptide bonds and links RNA and protein dominated worlds. *Nucleic Acids Res.*, 50, 1815–1828.

Català-Gorgues, J. (2022). The problem of design in the Darwinian proposal: A historical overview. In *Illuminating Human Evolution: 150 Years after Darwin*, Bertranpetit, J. and Peretó, J. (eds). Springer Nature, Singapore.

Collard, F., Baldin, F., Gerin, I., Bolsée, J., Noël, G., Graff, J., Veiga-da-Cunha, M., Stroobant, V., Vertommen, D., Houddane, A. et al. (2016). A conserved phosphatase destroys toxic glycolytic side products in mammals and yeast. *Nat. Chem. Biol.*, 12, 601–607.

Cornish-Bowden, A. and Cárdenas, M.L. (2020). Contrasting theories of life: Historical context, current theories. In search of an ideal theory. *Biosystems*, 188, 104063.

Cornish-Bowden, A., Cárdenas, M.L., Letelier, J.C., Soto-Andrade, J. (2007). Beyond reductionism: Metabolic circularity as a guiding vision for a real biology of systems. *Proteomics*, 7, 839–845.

Cornish-Bowden, A., Peretó, J., Cárdenas, M.L. (2014). Biochemistry and evolutionary biology: Two disciplines that need each other? *J. Biosci.*, 39, 13–27.

D'Ari, R. and Casadesús, J. (1998). Underground metabolism. *Bioessays*, 20, 181–186.

Darwin, C.R. (1859). *On the Origin of Species by Means of Natural Selection, or the Preservation of Favoured Races in the Struggle for Life*. John Murray, London.

Darwin, C.R. (1862a). Letter to Asa Gray, 23 July 1862. Letter 3662, Darwin correspondence project, University of Cambridge [Online]. Available at: https://www.darwinproject.ac.uk/letter/?docId=letters/DCP-LETT-3662.xml [Accessed 1 January 2023].

Darwin, C.R. (1862b). *On the Various Contrivances by which British and Foreign Orchids are Fertilised by Insects, and on the Good Effects of Intercrossing*. John Murray, London.

Darwin, C.R. (1874). *The Descent of Man, and Selection in Relation to Sex*, 2nd edition. John Murray, London.

Davidovich, C., Belousoff, M., Wekselman, I., Shapira, T., Krupkin, M., Zimmerman, E., Bashan, A., Yonath, A. (2010). The proto-ribosome: An ancient nano-machine for peptide bond formation. *Isr. J. Chem.*, 50, 29–35.

Dean, A.M. and Thornton, J.W. (2007). Mechanistic approaches to the study of evolution: The functional synthesis. *Nat. Rev. Genet.*, 8, 675–688.

Domínguez, M. (2017). *L'embolic de Darwin. Revisant la controvèrsia sobre l'ull en l'origen de les espècies*. Institut d'Estudis Catalans, Barcelona.

de Duve, C. (2005). *Singularities. Landmarks on the Pathways of Life*. Cambridge University Press, Cambridge.

Eck, R.V. and Dayhoff, M.O. (1966). Evolution of the structure of ferredoxin based on living relics of primitive amino acid sequences. *Science*, 152, 363–366.

Edge (2008). Engineering biology: A talk with Drew Endy [Online]. Available at: https://www.edge.org/conversation/drew_endy-engineering-biology [Accessed 20 December 2022].

Fischer, E. (1894). Einfluss der configuration auf die wirkung der enzyme. *Ber. Deutsch Chem. Gesellschaft*, 27, 2985–2993.

Fishman, R.S. (2010). Darwin and Helmholtz on imperfections of the eye. *Arch. Ophthalmol.*, 128, 1209–1211.

Fried, S.D., Fujishima, K., Makarov, M., Cherepashuk, I., Hlouchova, K. (2022). Peptides before and during the nucleotide world: An origins story emphasizing cooperation between proteins and nucleic acids. *J. R. Soc. Interface*, 19, 20210641.

Ghiselin, M.T. (1969). *The Triumph of the Darwinian Method*. University of California Press, Berkeley.

Hanczyc, M.M. and Monnard, P.A. (2017). Primordial membranes: More than simple container boundaries. *Curr. Opin. Chem. Biol.*, 40, 78–86.

Hanson, A.D., Henry, C.S., Fiehn, O., de Crécy-Lagard, V. (2016). Metabolite damage and metabolite damage control in plants. *Annu. Rev. Plant Biol.*, 67, 131–152.

Jabłońska, J. and Tawfik, D.S. (2021). The evolution of oxygen-utilizing enzymes suggests early biosphere oxygenation. *Nat. Ecol. Evol.*, 5, 442–448.

Jabłońska, J. and Tawfik, D.S. (2022). Innovation and tinkering in the evolution of oxidases. *Prot. Sci.*, 31, e4310.

Jacob, F. (1977). Evolution and tinkering. *Science*, 196, 1161–1166.

Jacob, F. (1981). *Le jeu des possibles. Essai sur la diversité du vivant*. Librairie Arthème Fayard, Paris.

Jacob, F. (1983). Molecular tinkering in evolution. In *Evolution from Molecules to Men*, Bendall, D.S. (ed.). Cambridge University Press, Cambridge.

Jacob, F. (2001). Complexity and tinkering. *Ann. New York Acad. Sci.*, 929, 71–73.

James, L.C. and Twafik, D.S. (2003). Conformational diversity and protein evolution: A 60-year-old hypothesis revisited. *Trends Biochem. Sci.*, 28, 361–368.

Jeffery, C.J. (2009). Moonlighting proteins: An update. *Mol. Biosyst.*, 5, 345–350.

Jensen, R.A. (1976). Enzyme recruitment in evolution of new function. *Annu. Rev. Microbiol.*, 30, 409–425.

Kacser, H. and Beeby, R. (1984). Evolution of catalytic proteins or on the origin of enzyme species by means of natural selection. *J. Mol. Evol.*, 20, 38–51.

Khersonsky, O. and Tawfik, D.S. (2010). Enzyme promiscuity: A mechanistic and evolutionary perspective. *Annu. Rev. Biochem.*, 79, 471–505.

Kim, J. and Copley, S.D. (2007). Why metabolic enzymes are essential or nonessential for growth of *Escherichia coli* K12 on glucose. *Biochemistry*, 46, 12501–12511.

Kim, J., Kershner, J.P., Novikov, Y., Shoemaker, R.K., Copley, S.D. (2010). Three serendipitous pathways in *E. coli* can bypass a block in pyridoxal-5'-phosphate synthesis. *Mol. Syst. Biol.*, 6, 436.

Krishnamurthy, R. (2020). Systems chemistry in the chemical origins of life: The 18th camel paradigm. *J. Syst. Chem.*, 8, 40–62.

Laurino, P., Tóth-Petróczy, Á., Meana-Pañeda, R., Lin, W., Truhlar, D.G., Tawfik, D.S. (2016). An ancient fingerprint indicates the common ancestry of Rossmann-fold enzymes utilizing different ribose-based cofactors. *PLoS Biol.*, 14, e1002396.

Lazcano, A. and Miller, S.L. (1999). On the origin of metabolic pathways. *J. Mol. Evol.*, 49, 424–431.

de Lorenzo, V. (2018). Evolutionary tinkering vs. rational engineering in the times of synthetic biology. *Life Sci. Soc. Pol.*, 14, 1–16.

Luisi, P.L. (2014). Prebiotic metabolic networks? *Mol. Syst. Biol.*, 10, 729.

Morange, M. (2002). Introduction. In *Travaux scientifiques de François Jacob*, Peyrieras, N. and Morange, M. (eds). Éditions Odile Jacob, Paris.

Morange, M. (2010). The resurrection of life. *Orig. Life Evol. Biosph.*, 40, 179–182.

Morange, M. (2017). What history tell us XLII. A "new" view of proteins. *J. Biosci.*, 42, 11–14.

Nam, H., Lewis, N.E., Lerman, J.A., Lee, D.H., Chang, R.L., Kim, D., Palsson, B.O. (2012). Network context and selection in the evolution to enzyme specificity. *Science*, 337, 1101–1104.

Nicholson, D.J. (2014). The machine conception of the organism in development and evolution. A critical analysis. *Stud. Hist. Phil. Sci. Part C: Stud. Hist. Phil. Biol. Biomed. Sci.*, 48, 162–174.

Nicholson, D.J. (2019). Is the cell really a machine? *J. Theor. Biol.*, 477, 108–126.

Noda-Garcia, L., Liebermeister, W., Tawfik, D.S. (2018). Metabolite-enzyme coevolution: From single enzymes to metabolic pathways and networks. *Annu. Rev. Biochem.*, 87, 187–216.

Noor, E., Flamholtz, A.I., Jayaraman, V., Ross, B.L., Cohen, Y., Patrick, W.M., Gruic-Sovulj, I., Tawfik, D.S. (2022). Uniform binding and negative catalysis at the origin of enzymes. *Prot. Sci.*, 31, e4381.

Notebaart, R.A., Szappanos, B., Kintses, B., Pál, F., Györkei, Á., Bogos, B., Lázár, V., Spohn, R., Csörgő, B., Wagner, A. et al. (2014). Network-level architecture and the evolutionary potential of underground metabolism. *Proc. Natl. Acad. Sci.*, 111, 11762–11767.

Orgel, L.E. (2000). Self-organizing biochemical cycles. *Proc. Natl. Acad. Sci.*, 97, 12503–12507.

Patel, B.H., Percivalle, C., Ritson, D.J., Duffy, C.D., Sutherland, J.D. (2015). Common origins of RNA, protein and lipid precursors in a cyanosulfidic protometabolism. *Nat. Chem.*, 7, 301–307.

Perchat, N., Dubois, C., Mor-Gautier, R., Duquesne, S., Lechaplais, C., Roche, D., Fouteau, S., Darii, E., Perret, A. (2022). Characterization of a novel β-alanine biosynthetic pathway consisting of promiscuous metabolic enzymes. *J. Biol. Chem.*, 298, 102067.

Peretó, J. (2012). Out of fuzzy chemistry: From prebiotic chemistry to metabolic networks. *Chem. Soc. Rev.*, 41, 5394–5403.

Peretó, J. (2019). Prebiotic chemistry that led to life. In *Handbook of Astrobiology*, Kolb, V. (ed.). CRC Press, Boca Raton.

Peretó, J. (2021). Transmetabolism: The non-conformist approach to biotechnology. *Microb. Biotechnol.*, 14, 41–44.

Piatigorsky, J., O'Brien, W.E., Norman, B.L., Kalumuck, K., Wistow, G.J., Borras, T., Nickerson, J.M., Wawrousek, E.F. (1988). Gene sharing by delta-crystallin and argininosuccinate lyase. *Proc. Natl. Acad. Sci.*, 85, 3479–3483.

Porcar, M. (2019). The hidden charm of life. *Life*, 9, 5.

Porcar, M. and Peretó, J. (2016). Nature versus design: Synthetic biology or how to build a biological non-machine. *Integr. Biol.*, 8, 451–455.

Pulletikurti, S., Yadav, M., Springsteen, G., Krishnamurthy, R. (2022). Prebiotic synthesis of α-amino acids and orotate from α-ketoacids potentiates transition to extant metabolic pathways. *Nat. Chem.*, 14, 1142–1150.

Romero Romero, M.L., Rabin, A., Tawfik, D.S. (2016). Functional proteins from short peptides. Dayhoff's hypothesis turns 50. *Angew. Chem. Int. Ed. Engl.*, 55, 15966–15971.

Rosen, R. (1991). *Life Itself: A Comprehensive Inquiry into the Nature, Origin, and Fabrication of Life*. Columbia University Press, New York.

Schwartz, A.W. (2007). Intractable mixtures and the origin of life. *Chem. Biodiver.*, 4, 656–664.

Springsteen, G., Yerabolu, J.R., Nelson, J., Rhea, C.J., Krishnamurthy, R. (2018). Linked cycles of oxidative decarboxylation of glyoxylate as protometabolic analogs of the citric acid cycle. *Nat. Commun.*, 9, 91.

Stubbs, R.T., Yadav, M., Krishnamurthy, R., Springsteen, G. (2020). A plausible metal-free ancestral analogue of the Krebs cycle composed entirely of α-ketoacids. *Nat. Chem.*, 12, 1016–1022.

Tawfik, D.S. (2010). Messy biology and the origin of evolutionary innovations. *Nat. Chem. Biol.*, 6, 692–696.

Tawfik, D.S. (2014). Accuracy-rate tradeoffs: How do enzymes meet demands of selectivity and catalytic efficiency? *Curr. Opin. Chem. Biol.*, 21, 73–80.

Valenti, R., Jabłońska, J., Tawfik, D.S. (2022). Characterization of ancestral Fe/Mn superoxide dismutases indicates their cambialistic origin. *Prot. Sci.*, 31, e4423.

Waley, S.G. (1969). Some aspects of the evolution of metabolic pathways. *Comp. Biochem. Physiol.*, 30, 1–11.

Xavier, J.C. and Kauffman, S. (2022). Small-molecule autocatalytic networks are universal metabolic fossils. *Phil. Trans. R. Soc. A*, 380, 20210244.

Xavier, J.C., Hordijk, W., Kauffman, S., Steel, M., Martin, W.F. (2020). Autocatalytic chemical networks at the origin of metabolism. *Proc. R. Soc. B*, 287, 20192377.

Yčas, M. (1974). On earlier states of the biochemical system. *J. Theor. Biol.*, 44, 145–160.

5

Viruses, Viroids and the Origins of Life

David DEAMER[1] and Marie-Christine MAUREL[2]

[1] Biomolecular Engineering, University of California, Santa Cruz, USA
[2] Institut de Systématique, Évolution, Biodiversité (ISYEB), École Pratique des Hautes Études, Muséum National d'Histoire Naturelle, Sorbonne Université, Université des Antilles, Paris, France

Viral diversity is immense and the virosphere is the largest reservoir of genetic diversity on Earth. Because 8% of the human genome consists of retroviral origin, we can say that they represent molecular "fossils" from ancient infections, hence their role in human evolution. In addition, subviral particles called viroids may be descendants of proviroids and are possibly remnants of an ancient RNA world. Many enigmas remain but we present here experimental results on the first RNA-like molecules in the form of rings that emerge in simulations of fresh water hydrothermal fields.

Over the past two millennia, pathogenic organisms have altered the course of human history time and time again. The Black Death pandemic, a bacterial plague, killed one-third of the population of Europe when it peaked between 1347 to 1351. Since 1983, the human immune deficiency virus (HIV) has resulted in 32 million deaths globally. The recent pandemic caused by a coronavirus is only the most recent example and has taught us how vulnerable we still are to viral infections. Once again, a tiny bit of nucleic acid – the SARS CoV-2 viral genome – has disrupted the normal course of civilization (Saleh and Rey 2021). We have been forcibly educated about the coronavirus with the image of a tiny sphere with spikes pointing outward, bearing a striking resemblance to a World War II naval mine. And like a mine, when a virus particle encounters a living cell and the spikes touch the membrane, it is internalized, it reproduces and the cell explodes, distributing hundreds of new infective particles into the surrounding medium.

But where did SARS CoV-2 come from? There are two answers to that question. The first concerns something that occurred in recent years when one of many coronaviruses circulating in wild animal populations happened to infect a human. The other answer we will explore here is that viruses have circulated in the biosphere ever since life began on the Earth.

5.1. How were viruses discovered? A brief history

We will begin with a quote from Isaac Asimov: "the most exciting phrase to hear in science, the one that heralds new discoveries, is not 'Eureka!' but 'that's funny...'".

Our understanding of viruses began in the late 1800s, not with a human disease but instead in tobacco plants that were grown commercially (Iwanowski 1942). In 1887, Iwanowski was a 23-year old graduate student in St. Petersburg, Russia. As part of his studies, he was sent to Ukraine and then to Crimea to investigate a strange disease that caused tobacco leaves to become mottled in a mosaic pattern. From an earlier publication by Adolf Mayer, who had studied the same disease in the Netherlands, he knew that it could be transmitted if the sap of a diseased plant was painted on leaves of healthy plants (Lustig and Levine 1992). Twenty years earlier, Louis Pasteur had established that bacteria were responsible for spoiling wine and beer. Ivanovsky knew this and decided to see whether the mottling was caused by bacteria. He put the sap of infected tobacco plants through a filter that removes bacteria, but the sap was still infective. He probably thought to himself, "that's funny...".

A few years later, the Dutch scientist Martinus Beijerinck repeated Ivanovsky's filtering experiment and found that the infection was not caused by an unknown bacterial species but instead by something entirely new (Beijerinck 1942; Creager 2002). Like most scientists of that era, Beijerinck knew Latin and decided to use the Latin word for poison or venom. Viruses were named for the first time in his 1898 publication. Further progress was made in 1935 when Wendell Stanley discovered that the tobacco mosaic virus, abbreviated TMV, could be crystallized

(Stanley 1946). Not only that, but the crystals were still infective! The importance of TMV was recognized when Stanley became one of four Nobel Prize winners in Chemistry in 1946. The remarkable properties of TMV were taken a step further when Fraenkel-Conrat and Williams (1955) showed that TMV could be disassembled into a protein fraction and a nucleic acid fraction, neither of which was infective. But if the two fractions were mixed, they spontaneously reassembled into an infective virus.

Are viruses alive? Probably not, because they can only reproduce in living cells. Viruses seem to exist somewhere between the living and nonliving, just as the earliest stage of life was a transition from random mixtures of organic material to nonrandom systems of organic compounds that exhibit the properties of life: metabolism, growth by polymerization and reproduction. It is reasonable to think that viruses have always existed in parallel with living cells. The novel coronavirus, SARS-CoV-2, is just one tiny branch of an evolutionary tree that goes back in time at least 4 billion years and may even have been involved in the origin of life.

5.2. Viral diversity

As the concept of viruses slowly penetrated the scientific community, research diverged into three paths defined by the kinds of cells that were infected. Two of the hosts – plants and animals – were obvious because of the diseases that the viruses caused. The filtration test devised by Ivanovsky became an important way to exclude bacteria as the causative agent of diseases. For instance, the infectious agent of foot and mouth disease in cattle passed this test and was the first animal virus reported in 1897, followed closely by the yellow fever virus in 1900 (Reed 1902). The third path was less obvious because the host cells are microscopic bacteria. We now know that an invisible battle is being waged continuously between bacteria and the viruses that infect them. The first hint was reported in 1896 by Ernest Hankin who made the surprising observation that an invisible agent in river water from the Ganges in India could destroy the bacteria responsible for cholera (Hankin 1896). Ten years later, Felix d'Herelle observed something that killed dysentery bacteria growing in Petri dishes. d'Herelle (1917, 1931) realized that this was a virus and invented the word bacteriophage[1], derived from Greek words for "bacteria eater".

Viral diversity is immense (Figure 5.1). Several hundred thousand species compose a special niche within the biosphere called the virosphere, which is the largest reservoir of genetic diversity on Earth.

---

1 For over 95% of known phages, the genetic material is a double-stranded DNA molecule between 5 and 650 kpb in length.

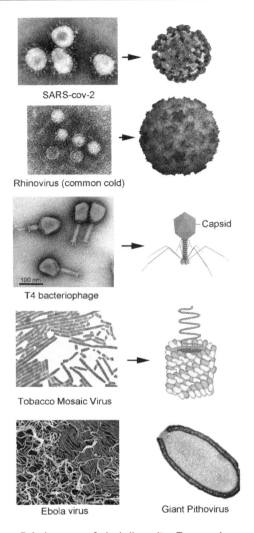

Figure 5.1. *Images of viral diversity. For a color version of this figure, see www.iste.co.uk/dimauro/firststeps.zip*

NOTE ON FIGURE 5.1.– *Viruses are too small to be seen with light microscopes which can only show 1,000× magnification. The electron microscope has much more resolving power and electron micrographs of four different viruses are on the left with artists illustrations on the right. The bottom row shows electron micrographs of the ebola and giant pithovirus (pithovirus image credit Julia Bartoli, Chantal Abergel, CNRS).*

Although most of our attention is directed toward those relatively few viruses that cause disease, in terms of numbers, virions (viral genome surrounded by a capsid) dominate the biomass of life on Earth. For instance, ~10^{30} virions are in the ocean, equivalent to billions per liter of seawater. The adult human body has approximately 37 trillion cells, yet it is estimated that there are 100 times more viruses than cells, mostly coexisting with bacteria in the gut. Although viruses have traditionally been classified outside the tree of life, the distinction is not clear and they are proving to be much more complex than originally thought. Recently discovered DNA viruses approach bacteria in size and complexity. Claverie and Abergel (2010) in Marseille have discovered and characterized giant mimiviruses (mimicking microbe virus) that are larger than some bacteria and have a genome composed of a thousand genes. Even larger viruses infect amoebas and they themselves are infected by a small virus called Sputnik (La Scola et al. 2008).

5.3. Viral structure and function

Viruses are generally submicroscopic, ranging from 20 to 300 nm. Poliovirus is one of the smallest viruses, 28 nm in diameter. To give a perspective on these dimensions, the size of a glucose molecule is approximately 1 nm and hemoglobin is 5 nm across. The largest known virus so far is the Pithovirus that is isolated from the permafrost of Siberia (Legendre et al. 2014). Pithovirus is 1.5 µm long, the size of a typical bacterial cell, and was still active after 32,000 years frozen in icy soil.

Viruses have an extracellular state referred to as virions, which contain the genetic material within compartments called capsids. Most virions are fixed in size, but there is at least one exception to the rule. Prangishvili et al. (2006, 2018) studied archaea viruses called ATV (Acidianus two-tailed virus). Bicaudaviridae is a family of hyperthermophilic archaeal viruses. Members of the genus *Acidianus* serve as natural hosts. There is only one genus (*Bicaudavirus*) and one species in this family, that is, ATV.

The virions are lemon-shaped when first released but then grow appendages from both ends until they resemble a spindle. ATV lack ribosomes and metabolism, so the appendages are presumably produced from substances stored in the capsid rather than being synthesized.

While all living cells contain functional DNA and RNA, each virus species uses only one of the two varieties of nucleic acids. For instance, the herpes virus, adenoviruses, bacteriophages and hepatitis B virus have a DNA genome. Other viruses use RNA for their genome, such as the coronavirus SARS-CoV-2 responsible for Covid-19, which has an RNA genome 30,000 nucleotides in length that codes for

29 proteins. Another group of RNA viruses are called retroviruses. Examples include the poliomyelitis virus and the HIV (human immunodeficiency virus). Some bacteriophages infect bacterial cells by using a set of leg-like proteins to bind to the surface of the host bacterium and then literally injecting its DNA genome into the cell (Figure 5.2). The DNA is reproduced by enzymes called polymerases, and the genes in the phage DNA are transcribed into messenger RNA which then directs the synthesis of capsid proteins by ribosomes. The DNA and proteins assemble into mature virions that are released when the cell dies (Alberts et al. 2002).

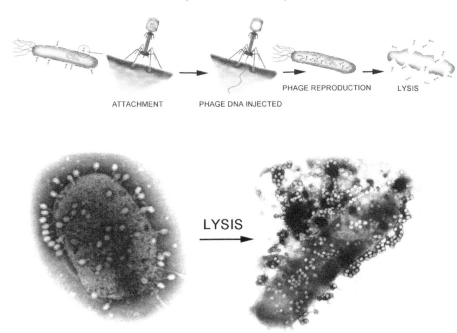

Figure 5.2. *Infection of a bacterial cell by a bacteriophage followed by lysis. Biozentrum, University of Basel. For a color version of this figure, see www.iste.co.uk/dimauro/firststeps.zip*

Infection by a virus with an RNA genome such as the coronavirus is more complicated. In a first step, virion capsid proteins adhere to receptors which are molecules located on the cell surface (Figure 5.3). For instance, in SARS-CoV-2, the spike proteins that form the crown of the coronavirus virion are composed of two subunits that bind the virion to ACE2, the target receptor on the cell surface (Jackson et al. 2022). One of the subunits mediates fusion of the viral envelope with the cell membrane and the RNA is released inside. The RNA then initiates two actions. The first is simply to be repeatedly transcribed and replicated by a polymerase enzyme that was synthesized from one of the genes in the viral RNA. The new copies serve as the equivalent of messenger RNA by taking over ribosomes and guiding the synthesis of the viral proteins composing the capsid. As viral RNA and capsid proteins accumulate in the cell, they spontaneously assemble into complete virions. The process can be surprisingly fast, with a hundred or more virions assembling in 20 minutes. Ultimately the cell bursts and the virions are released, ready to infect neighboring cells.

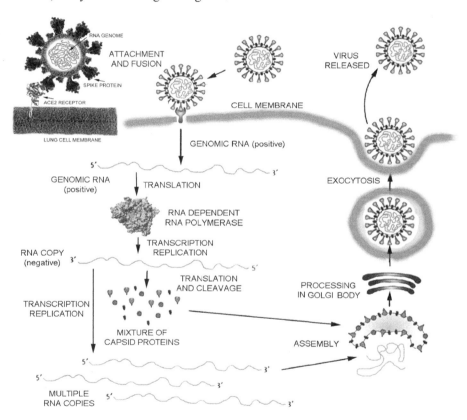

Figure 5.3. *Diagram of the SARS coronavirus reproduction. The protein capsid is a vehicle that binds to a cell membrane, then fuses to release the RNA into the cell. The RNA is 30,000 nucleotide monomers in length and contains enough genetic information to guide the synthesis of 29 proteins. Illustration adapted from public domain images. For a color version of this figure, see www.iste.co.uk/dimauro/firststeps.zip*

In some cases, a kind of budding occurs and the cell is not lysed but continues to store and produce virions. The infection can be latent, which means that there is a delay between the time of infection and the appearance of symptoms. Fever blisters caused by the herpes virus are typical of latent viral infection, and symptoms reappear from time to time when the virus emerges from its "latency".

Another example is shingles in adults which is caused by the chickenpox virus (variola) (Hope-Simpson 1965). The infection occurs in childhood and then years pass before the virus reappears as a painful rash. A few viral infections can also cause normal tissues to undergo transformation into cancers. A prominent example is cervical cancer caused by the papilloma virus, which can be prevented by the HPV vaccine administered during adolescence (Markowitz et al. 2018).

5.4. Viruses and mammalian genomes

An astonishing 8% of the human genome consists of sequences of retroviral origin (Zahn et al. 2015). These represent molecular "fossils" remaining from ancient infections tens of millions of years ago. Because they have undergone mutations over many generations, most of the sequences are degenerate and inactive but a few remain intact despite the time that has passed since the original retroviral infection. It is an amazing fact that humans are placental mammals because of RNA viruses. Sequences called HERVs (Human Endogenous RetroVirus) represent remnants of RNA viruses that somehow became integrated into the mammalian genome. These genes are found in many placental mammals, including primates, rodents, carnivores and ruminants, presumably from viruses that infected mammals several times between 70 and 25 million years ago. One of the HERV sequences directs the synthesis of a protein called Syncytin-1, which induces cell fusion events between the cytotrophoblastic cells of the placenta, allowing the formation of the cell structure named syncytiotrophoblast (Mi 2000). Syncytins have a second property: like the envelope proteins of infectious retroviruses, they can inhibit defensive functions of the immune system. This immunosuppressive property plays a central role in the interaction between the fetus and the mother during gestation and accounts for the fact that even though half of the fetal proteins are paternal and therefore potentially antigenic, the fetus is not rejected by the mother's immune system.

We conclude from these remarkable results that the evolutionary acquisition of viviparity could only occur in parallel with multiple introductions of viral genomes into mammalian genomes. This phenomenon continues today. For example, Australian koalas (marsupials) are subject to infection by the koala retrovirus (KoRV) (Stoye 2006). Since their numerous genes have no homology with the three domains of life, an independent origin is therefore conceivable, hinting at the existence of a fourth branch in the tree of life.

5.5. Role of viruses in human evolution, health and disease

Although the evolutionary role of viruses is fascinating, what is most important to humans today is the vast range of diseases they cause (Cagliani et al. 2021). Virtually every tissue and organ are the target of one or several viral species. Viral epidemics are mainly zoonoses: diseases produced by the transmission of a pathogen from animals to humans. Recent examples include HIV from chimpanzees, Ebola in Africa which has its source in "bush meat" from wild animals such as bats, varieties of influenza like swine flu and bird flu, and most recently the novel coronavirus from bats and pangolins (Andersen 2020). The increasingly frequent emergence of these infectious diseases corresponds to our growing invasion of natural environments as the human population grows. Deforestation is taking place and wild animals being hunted in their natural habitat are brought into close contact with human populations. The result is that novel previously unknown viruses are transmitted into new host organisms, in this case human beings.

Because of the current prominence of Covid-19, it is worth taking a moment to understand some of the molecular details involved in the mechanism by which it was transmitted to humans. We can begin with the human version and step backward in time. SARS-CoV-2 proteins are synthesized in the host cell under the direction of the genes in the viral RNA, but the protein needs to be broken into smaller pieces before they can function, a process called cleavage. There is a cleavage site in SARS-CoV-2 that has an amino acid called proline inserted, and this seems to increase infectivity because it undergoes cleavage at a greater rate.

How did CoV-2-SARS acquire this unique cleavage site that is not present in pangolin coronavirus? Either SARS-CoV-2 acquired this characteristic in an animal host, or it acquired it in humans after transmission of the pangolin coronavirus to humans. It is known that some viruses eventually acquire a cleavage site spontaneously. For instance, other human coronaviruses such as HCoV-HKU1 have a similar cleavage site. It is therefore possible that mixing of a human coronavirus with the pangolin coronavirus may have resulted in SARS-CoV-2. The hypothesis of a primitive pangolin coronavirus circulating widely, silently and for a long time in people involved in the bushmeat wildlife trade is therefore plausible. The AIDS epidemic has a similar trajectory of viral mixing within primate populations followed by transmission to hundreds of millions of people.

5.6. Viroids may be a link to ancient evolutionary pathways

Potato spindle tuber disease was described in 1922 (Schultz and Folsom 1923). The disease was first attributed to bacteria, then to viruses, but Theodor Otto Diener, a plant disease specialist, reported that the pathogen apparently was a form of free RNA which he called viroid (Diener 1971). At first, the existence of infectious naked RNA was doubted, but experimental infections were carried out that revealed pathogenic RNAs smaller than the smallest viruses were responsible for devastating plant diseases. When a plant is infected with a viroid, most of the symptoms occur in the tissues and organs. Foliage and stems may be deformed, discolored or otherwise damaged. Flowers are necrotic, small and depigmented. Fruit size, shape and quality are also affected.

There are 34 known species of viroids divided into two families: the Pospiviroidae and the Avsunviroidae. It is remarkable that Avsunviroidae replicate in the green chloroplasts of plant cells that are descendents of cyanobacteria, the first bacteria capable of photosynthesis, while Pospiviroidae replicate in the cell nucleus (Góra-Sochacka 2004). Viroids, together with satellite RNAs, are the only RNA molecules known to be able to replicate without DNA intermediates. They do not code for any protein and are dependent on host enzymes for their replication.

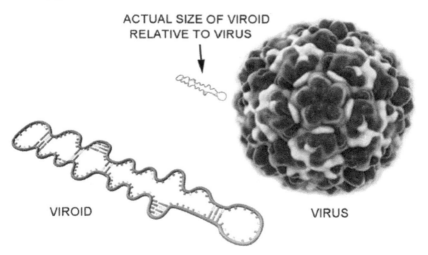

Figure 5.4. *Structure of a viroid compared to a viral particle. Viroids are the smallest known infective agents and are composed of just a few hundred nucleotides that form a ring. They are much smaller than a typical virus particle. Illustration adapted from public domain images*

Viroids lack an envelope or capsid and are little more than a single strand of circular RNA consisting of 246–401 ribonucleotides (Figure 5.4). Unlike viruses, they have no known cellular receptors. Infection and transmission have been well studied in the case of coconut cadang cadang viroid and coconut tinangaja viroid, which infect palms and coconut palms (Haseloff et al. 1982). To enter the plant cells, they must somehow get through the cell wall and cell membrane which serve to protect the delicate intracellular machinery (Ding et al. 2005). The viroids do this by finding their way through openings caused by insect bites or wounds produced by pruning with agricultural tools. Once they begin to replicate in a cell, the viroids move from cell to cell through the phloem, or circulatory system of the plant. Death of the plants may occur, leading to economic disasters.

Not all viroids are pathogenic. Some are asymptomatic, such as the hop viroid, and some are even beneficial to their hosts. The production of dwarf lemons after infection by asymptomatic viroids has been developed in plantations in Dareton (Australia) to allow an increase in planting density for a better harvest. Yields per planted hectare were found to be higher for viroid-inoculated trees compared to the same non-inoculated trees (Hutton et al. 2000).

To date, all known viroids have been discovered in plants, but their different sites of replication (nucleus or chloroplast) as well as the presence of viroid-like domains in the RNA of the human HDV (hepatitis delta virus) reveal their diversity and ability to adapt to different environments. Viroid and HDV RNAs share several characteristics, including circular structure, compact folding and replication via a rolling circle mechanism (Diener 1987). For instance, the avocado sunblotch viroid is capable of replicating and maintaining itself for at least 25 generations in the yeast *Saccharomyces cerevisiae* (Delan-Forino et al. 2011). Replication was also observed in a cyanobacterium called Nostoc (Latifi et al. 2016). It should be added that, although viroids have never been demonstrated in the animal world, three exceptionally stable small RNAs have been associated with Crohn's disease and ulcerative colitis (Roy et al. 1997).

5.7. Origin and evolution of viroids

The origin of viroids can be compared with what could have happened at the dawn of life (Diener 2001). By a yet unknown process, organic compounds on the prebiotic Earth underwent spontaneous self-assembly to form vast numbers of microscopic structures called protocells. These were not alive, but they were subject to selection. Over many millions of years, more robust protocells emerged and at some point, they contained catalytic polymers like RNA that were able to replicate. This concept became more plausible in the early 1980s when Tom Cech and Sidney Altman discovered RNA catalysts now called ribozymes (Grabowski et al. 1982; Guerrier-Takada 1983; Maurel et al. 2019). The hypothesis that RNA preceded DNA as a carrier of genetic information and a catalyst for early metabolic reactions became more plausible. RNA, both genotype and phenotype, would have initiated Darwinian evolution at the molecular level in the absence of DNA and proteins.

Although the evolutionary origin of viroids remains uncertain, it is possible that viroids are remnants of an ancient RNA world originally populated by free RNAs. These proviroids at some point in evolution entered and became dependent on cellular organisms.

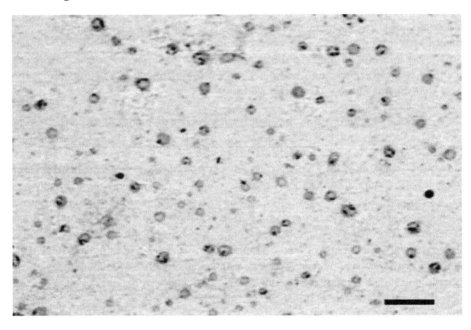

Figure 5.5. *RNA rings can be visualized by atomic force microscopy. The rings form spontaneously when monomer nucleotides of RNA are exposed to multiple cycles of wetting and drying that simulate conditions on the prebiotic Earth (Hassenkam et al. 2020). Scale bar: 200 nm*

One hypothesis is that current viroids may be descendants of proviroids that invaded ancient cyanobacteria (Weia et al. 2019; Moelling and Broecker 2021). These organisms became endosymbiotes of primitive plants and evolved into chloroplasts. Once inside the chloroplasts, some viroids could eventually move to the nucleus and become dependent on nuclear polymerases. However, the idea that there would be a viroid branch in the tree of life is not unanimously accepted. The debate is lively because the concept of species as the basic unit of taxonomy is still vague and poorly defined.

But what was the source of the first RNA-like molecules? The authors have addressed this question and several publications have demonstrated that RNA-like polymers can be synthesized by cycles of wetting and drying that simulate the conditions of hydrothermal freshwater fields associated with volcanic landmasses on the early Earth (Rajamani et al. 2008; Da Silva et al. 2014; DeGuzman et al. 2014; Damer and Deamer 2020). Most recently, we have reported evidence from atomic force microscopy that some of the RNA synthesized under these conditions takes the form of circular rings composed of several hundred nucleotides, the same dimensions as viroids (Hassenkam et al. 2020). An interesting question for future research is whether such rings have the unexpected capability of growing by a template-directed addition of mononucleotides.

5.8. Conclusion

Ever since primitive forms of life emerged, the biological world has undergone continuous incremental evolution toward ever increasing complexity. Life presumably began when physical and chemical properties allowed diverse organic compounds to assemble into structures capable of undergoing selection and then evolution. It is likely that a variety of nucleoside monomers were synthesized by nonenzymatic reactions on the early Earth; so the eight canonical species that became dominant in life today are the products of an evolutionary process. More than 30 noncanonical nucleosides have been discovered in all three domains of life (Schneider et al. 2018). These may represent remnants of a time when the molecular machinery of life had yet to be refined by several billion years of evolution.

The origin of viroids remains an enigma, but it is possible that they are relics of a time preceding cellular life when noncoding RNA molecules catalyzed their own synthesis (Kristoffersen 2022). If so, ancestral viroids would have served as the foundation of an RNA world that preceded the current world based on DNA, RNA and proteins. In the course of evolution, these free molecules survived by acquiring an intracellular lifestyle in which abundant nutrients were available. The hypothesis that viroids preceded an RNA world is an exciting question (Diener 1989) to be studied experimentally and phylogenetically to try to understand how they are related to the origin and evolution of life on Earth. Although recent phylogenetic analyses have shown that viral RNAs are the probable ancestors of DNA viruses, technical and conceptual challenges must still be overcome to advance to a deeper understanding of viroid origin and evolution.

5.9. References

Alberts, B., Johnson, A., Lewis, J., Raff, M., Roberts, K., Walter, P. (2002). The life cycle of bacteriophage lambda. In *Molecular Biology of the Cell*, 4th edition. Garland Science, New York.

Andersen, K.G., Rambaut, A., Lipkin, W.I., Holmes E.C., Garry R.F. (2020). The proximal origin of SARS-CoV-2. *Nature Medicine*, 26, 450–455.

Beijerinck, M.W. (1942). Concerning a Contagium vivum fluidum as cause of the spot disease of tobacco leaves, translated by James Johnson. In *Phytopathology Classic, Number 7*, APS Publications, St. Paul, MN.

Cagliani, R., Mozzi, A., Pontremoli, C., Sironi, M. (2021). Evolution and origin of human viruses. In *Virology*, Saleh, M.C. and Rey, F.A. (eds). ISTE Ltd, London and John Wiley & Sons, New York.

Claverie, J.M. and Abergel, C. (2010). Mimivirus: The emerging paradox of quasi-autonomous viruses. *Trends Genet.*, 26, 431–437.

Creager, A.N.H. (2002). *The Life of a Virus: Tobacco Mosaic Virus as an Experimental Model, 1930–1965*. University of Chicago Press, Chicago.

Da Silva, L., Maurel, M.C., Deamer, D. (2014). Salt-promoted synthesis of RNA-like molecules in simulated hydrothermal conditions. *J. Mol. Evol.*, 80, 86–97.

Damer, B. and Deamer, D. (2020). The hot spring hypothesis for an origin of life. *Astrobiology*, 20, 429–452.

De Guzman, V., Shenasa, H., Vercoutere, W., Deamer, D. (2014). Generation of oligonucleotides under hydrothermal conditions by non-enzymatic polymerization. *J. Mol. Evol.*, 78, 251–262.

Delan-Forino, C., Maurel, M.-C., Torchet, C. (2011). Replication of *avocado sunblotch viroid* in the yeast *Saccharomyces cerevisiae*. *Journal of Virology*, 85, 3229–3238.

Diener, T.O. (1971). Potato spindle tuber "virus" IV. A replicating, low molecular weight RNA. *Virology*, 45, 411–428.

Diener, T.O. (ed.) (1987). *The Viroids*. Plenum Press, New York.

Diener, T.O. (1989). Circular RNAs: Relics of precellular evolution? *Proc. Natl. Acad. Sci. USA*, 86, 9370–9374.

Diener, T.O. (2001). The viroid: Biological oddity or evolutionary fossil? *Adv. Virus Res.*, 57, 137–184.

Ding, B., Asuka Itaya, A., Xuehua, Z. (2005). Viroid trafficking: A small RNA makes a big move. *Current Opinion in Plant Biology*, 8(6), 606–612.

Fraenkel-Conrat, H. and Williams, R.C. (1955). Reconstitution of active tobacco mosaic virus from its inactive protein and nucleic acid component. *Proc. Natl. Acad. Sci. USA*, 41, 690–698.

Góra-Sochacka, A. (2004). Viroids: Unusual small pathogenic RNAs. *Acta Biochimica Polonica*, 51, 587–607.

Guerrier-Takada, C., Gardiner, K., Marsh, T., Pace, N., Altman, S. (1983). The RNA moiety of ribonuclease P is the catalytic subunit of the enzyme. *Cell*, 35, 849–857.

Hankin, E.H. (1896). L'action bactéricide des eaux de la Jumna et du Gange sur le vibrion du choléra. *Annales de l'Institut Pasteur*, 10, 511–523.

Haseloff, J., Mohamed, N.A., Symons, R.H. (1982). Viroid RNAs of cadang-cadang disease of coconuts. *Nature*, 299, 316–321.

Hassenkam, T., Damer, B., Mednick, G., Deamer, D. (2020). Viroid-sized rings self-assemble from mononucleotides through wet-dry cycling: Implications for the origin of life. *Life (Basel)*, 10, 321–332.

d'Hérelle, F. (1917). Sur un microbe invisible antagoniste des bacilles dysentériques. Note de M.F. d'Hérelle, présentée par M. Roux. *Comptes rendus de l'académie des sciences*, 165, 373–375.

d'Herelle, F. (1931). An address on bacteriophagy and recovery from infectious diseases. *Canadian Medical Association Journal*, 24, 619–628.

Hope-Simpson, R.E. (1965). The nature of herpes zoster: A long-term study and a new hypothesis. *Proc. R. Soc. Med.*, 58, 9–20.

Hutton, R.J., Broadbent, P., Bevington, K.B. (2000). Viroid dwarfing for high density citrus plantings. *Hort. Rev.*, 24, 277–331.

Iwanowski, D. (1942). Concerning the mosaic disease of tobacco, translated by James Johnson. In *Phytopathological Classics*, APS Publications, St. Paul, MN.

Jackson, C.B., Farzan, M., Chen, B., Choe, H. (2022). Mechanisms of SARS-CoV-2 entry into cells. *Nat. Rev. Mol. Cell Biol.* 23, 3–20.

Kristoffersen, E.L., Burman, M., Noy, A., Holliger, P. (2022). Rolling circle RNA synthesis catalyzed by RNA. *eLife*, 11, e75186.

Kruger, K., Grabowski, P.J., Zaug, A.J., Sands, J., Gottschling, D.E., Cech, T.R. (1982). Self-splicing RNA: Autoexcision and autocyclization of the ribosomal RNA intervening sequence of tetrahymena. *Cell*, 31, 147–157.

La Scola, B., Desnues, C., Pagnier, I., Robert, C., Barrassi, L., Fournous, G., Merchat, M., Suzan-Monti, M., Forterre, P., Koonin E., Raoult D. (2008). The virophage as a unique parasite of the giant mimivirus. *Nature*, 455, 100–104.

Latifi, A., Bernard, C., da Silva, L., Andéol, Y., Elleuch, A., Risoul, V., Vergne, J., Maurel, M.-C. (2016). Replication of avocado sunblotch viroid in the cyanobacterium *Nostoc* sp. PCC 7120. *J. Plant Pathol. Microbiol.*, 7, 341.

Legendre, M., Bartoli, J., Shmakova, L., Jeudy, S., Labadie, K., Adrait, A., Lescot, M., Poirot, O., Bertaux, L., Bruley, C. et al. (2014). Thirty-thousand-year-old distant relative of giant icosahedral DNA viruses with a pandoravirus morphology. *Proc. Natl. Acad. Sci. USA*, 11, 4274–4279.

Lustig, A. and Levine, A.J. (1992). One hundred years of virology. *American Society for Microbiology*, 66, 4629–4631.

Markowitz, L.E., Gee, J., Chesson, H., Stokley, S. (2018). Ten years of human papillomavirus vaccination in the United States. *Acad Pediatr.*, 18(2S), S3–S10.

Maurel, M.-C., Leclerc, F., Vergne, J., Zaccai, G. (2019). RNA back and forth: Looking through ribozyme and viroid motifs. *Viruses*, 11, 283–300.

Mi, S. (2000). Syncytin is a captive retroviral envelope protein involved in human placental morphogenesis. *Nature*, 403, 785–789.

Moelling, K. and Broecker, F. (2021). Viroids and the origin of life. *Int. J. Mol. Sci.*, 22, 3476.

Prangishvili, D., Vestergaard, G., Häring, M., Aramayo, R., Basta, T., Rachel, R., Garrett, R.A. (2006). Structural and genomic properties of the hyperthermophilic archaeal virus ATV with an extracellular stage of the reproductive cycle. *J. Mol. Biol.*, 359, 1203–1216.

Prangishvili, D., Krupovic, M., ICTV Report Consortium (2018). ICTV virus taxonomy profile: Bicaudaviridae. *The Journal of General Virology*, 99(7), 864–865.

Reed, W. (1902). Recent researches concerning the etiology, propagation, and prevention of yellow fever by the United States Army Commission. *J. Hyg.*, 2, 101–119.

Rajamani, S., Vlassov, A., Benner, S., Coombs, A., Olasagasti, F., Deamer, D. (2008). Lipid-assisted synthesis of RNA-like polymers from mononucleotides. *Orig. Life Evol. Biosph.*, 38, 57–74.

Saleh, M.-C. and Augusto Rey, F. (eds) (2021). Overview. In *Virology*. ISTE Ltd, London, and John Wiley & Sons, New York.

Schneider, C., Becker, S., Okamura, H., Crisp, A., Amatov, T., Stadlmeier, M., Carell, T. (2018). Noncanonical RNA nucleosides as molecular fossils of an early Earth-generation by prebiotic methylations and carbamoylations. *Angew. Chem. Int. Ed. Engl.*, 57, 5943–5946.

Schultz, E.S. and Folsom, D. (1923). Transmission, variation, and control of certain degeneration diseases of Irish potatoes. *J. Agric. Res.*, 25, 43.

Stanley, W.M. (1946). The isolation and properties of crystalline tobacco mosaic virus. Nobel Lecture, December 12, 1946.

Stoye, J.P. (2006). Koala retrovirus: A genome invasion in real time. *Genome Biology*, 7(11), 241.

Weia, S., Biana, R., Andikab, I.B., Niua, E., Liua, Q., Kondoc, H., Yanga, L., Zhoua, H., Panga, T., Liana, Z. et al. (2019). Symptomatic plant viroid infections in phytopathogenic fungi. *Proc. Natl. Acad. Sci. USA*, 116, 13042–13050.

Zahn, J., Kaplan, M.H., Fischer, S., Dai, M., Meng, F., Saha, A.K., Cervantes, P., Chan, S.M., Dube, D., Omenn, G.S., Markovitz, D.M., Contreras-Galindo, R. (2015). Expansion of a novel endogenous retrovirus throughout the pericentromeres of modern humans. *Genome Biol.*, 16, 74.

6

Is the Heterotrophic Theory of the Origin of Life Still Valid?

Antonio LAZCANO[1,2]

[1] Universidad Nacional Autónoma de México, Mexico City, Mexico
[2] El Colegio Nacional, Mexico City, Mexico

6.1. Introduction

"There is grandeur in this view of life, with its several powers, having been originally breathed by the Creator into a few forms or into one", wrote Darwin in the closing paragraph of the second edition of *The Origin of Species*. He ended by adding that "whilst this planet has gone cycling on according to the fixed law of gravity, from so simple a beginning endless forms most beautiful and most wonderful have been, and are being, evolved". Ernst Haeckel's strong loyalty to Darwin's ideas did not prevent him from rejecting a divine origin of life. As he wrote in a footnote in his monograph on the Radiolaria "The chief defect of the Darwinian theory is that it throws no light on the origin of the primitive organism—probably a simple cell—from which all the others have descended. When Darwin assumes a special creative act for this first species, he is not consistent, and, I think, not quite sincere..." (Haeckel 1862).

Haeckel's reference to "a simple cell" as the starting point of life reflects his deep commitment both to Darwinism and the cell theory, which he studied under Schleiden and Virchow. He consistently argued that the oldest forms of life were the Monera, which at the time were considered as nothing more than non-nucleated undifferentiated small globules of protoplasm. Although nowadays the term protoplasm has been confined to history or biology books, for 150 years it was in the limelight when the French microbiologist Felix Dujardin started crushing ciliates under the microscope, and observed that the tiny cells exuded a protein-rich jellylike substance, which he described as a "geleé vivante", that was eventually named "protoplasm" in the late 1830s by Johann E. Purkinje and Hugo von Mohl (Lazcano 2013).

Like many of his contemporaries, Haeckel assumed that the properties of protoplasm could be explained by those of proteins, which many believed were its basic components. A committed monist, Haeckel did not recognize any essential difference between inert and living matter. As he wrote in his 1900 book *The Riddle of the Universe*:

> [...] the chemical elements that exist in living beings are the same as those in the inorganic world [...] organic life is itself a chemical-physical process, based on the metabolism (or exchange of material) of these albuminates [...] These protoplasmic carbon compounds are distinguished from most other chemical combinations by their highly intricate molecular structure, their instability, and their gelatinous consistency.

Haeckel had an extraordinary panoramic vision of biological phenomena and may have been the first to go as far as to propose a scheme of what we now call cosmic evolution, which included the possibility of an evolutionary continuity between the non-living and the living through spontaneous generation.

6.2. The roaring 20s

Haeckel's proposal spread among Russian researchers and naturalists, whose strong scientific ties with Germany went back to the times of Catherine the Great (Kolchinsky and Levit 2019). Like many other Russian students, the young Alexandr I. Oparin rapidly became familiar with Haeckel's publications. As argued here, it is impossible to understand the origin and development of his ideas and some of the ensuing scientific and political debates without acknowledging the key role that Haeckel played in the development of Oparin's views on nature and the origin of life.

Oparin's formative stages took place in the midst of the turbulent years that marked the end of the Russian Empire, the rise to power of the Bolsheviks, and the ensuing Civil War. Historians have recorded the intense political and social changes of those times, but the most moving accounts of the individual and collective tragedies that took place during that period are found in gripping novels, including Pasternak's Dr. Zhivago and Bulgakov's White Guard. This was also a period of extraordinary creativity and intense social and cultural activity represented by various avant garde groups of painters, musicians, poets and visual artists, many of whom were quite independent from the dominant revolutionary groups. The same was not true for scientists, as many openly sided with the Bolsheviks. Those years also witnessed intense discussions on Darwinism, that since the late 19th century had become, often in opposition to Mendel's discoveries, a battle cry of the most liberal and revolutionary sectors of Russian society as part of their struggle against the conservative monarchist establishment.

The scientists who played a pivotal influence role in Oparin formative processes and in the development of his ideas were Aleksei N. Bakh and Kliment A. Tymiriazev, two Russian researchers who earned an international recognition for their academic work and whose infatuation with the Soviet scientific and political apparatus is well documented. Bakh's work on the assimilation of carbon dioxide by plants was based on the ideas of the German chemist Adolf von Baeyer, who argued that the laboratory formation of sugars from formaldehyde proved that this molecule was as an intermediary in photosynthesis. After spending more than 30 years abroad, Bakh returned to Moscow in 1917, befriended Lenin, and set up a laboratory that would later become the Karpov Physical Chemical Institute. Driven by his strong interest in biochemistry, the 23-years-old Oparin quickly joined Bakh's group and rapidly became familiar with the properties of enzymatic processes and the molecular study of photosynthesis.

Like Bakh, Tymiriazev had also worked on the chemistry of CO_2 and the properties of chlorophyll. He was a well-known agronomist, but his deep interest in evolution led him to visit Down House, and he very quickly became the main promoter of Darwin's ideas, first in the Russian Empire and later in the Soviet Union. Unlike Bakh, he did not go into exile, but he resigned from his teaching position at Moscow State University to protest against the repressive policies of the Imperial Minister of Education Lev A. Kasso. Many decades later, Oparin still remembered those times and how Timiryazev had continued his advocacy of evolutionary ideas by organizing weekly meetings with colleagues and some students in his Moscow apartment (Lazcano 1992). Thus, from the early stages of his scientific career, Oparin developed strong ties with the upper echelons of the Soviet scientific establishment.

Oparin's strong interest in Darwin's ideas rapidly caught Tymiriazev's attention, who invited him to attend the meetings held at his home. Thus, from the initial phases of his scientific career, Oparin had privileged access to up-to-date evolutionary literature and Darwinian analysis methodologies, while deepening his understanding of the biochemical mechanisms of metabolic pathways because of his association with Bakh. He soon realized that from an evolutionary perspective, the prevailing idea of a photosynthetic origin of life was not tenable, and concluded that the first living entities must have been heterotrophic anaerobic Monera. In his eyes, this possibility appeared to be confirmed by universality of fermentative metabolism, which in Darwinian terms implied that extant living beings descended from an anaerobic heterotrophic ancestor.

If the first organisms had been anaerobic heterotrophs, then a source of carbon and energy was required. Based on an in-depth review of the chemical composition of meteorites and the presence of HCN and CH+ cometary spectra, on the one hand, and the rich 19th-century tradition of organic chemistry, on the other hand, which achieved the laboratory synthesis of urea, hydrocarbons, sugars, amino acids and other organic compounds of biochemical significance, Oparin argued that the primitive Earth was an anoxic environment that led to the synthesis and accumulation of systems of organic molecules that gave rise to the first organisms. He thus not only developed a step-wise evolutionary proposal of the events that preceded the first living beings, but also established the multi- and interdisciplinary research program that continues to characterize studies on the emergence of life to this day.

In 1924, Oparin published his theory in a small book titled *The Origin of Life*. At least three additional proposals of a heterotrophic origin of life appeared during those years. They were published by the microbiologist R.B. Harvey, the geochemist Charles Lipman, and the brilliant British polymath John B.S. Haldane (Bada and Lazcano 2003), who coined the term "hot dilute soup", a metaphor that continues to be used (and misinterpreted) to this day. The issue of plagiarism cannot be raised. As Francis Galton wrote in his 1869 book:

> It is notorious that the same discovery is frequently made simultaneously and quite independently, by different persons [...] It would seem, that discoveries are usually made when the time is ripe for them—that is to say, when the ideas from which they naturally flow are fermenting in the minds of many men. When apples are ripe, a trifling event suffices to decide, which of them shall first drop off its stalk, so a small accident will often determine the scientific man who shall first make and publish a new discovery.

6.3. Coacervates as models of precellular structures

Oparin's original proposal represented a major breach with his scientific mentors and predecessors who had not only advocated spontaneous generation, but had also stressed the autotrophic nature of the first forms of life. Instead, what Oparin did was explain the origin of the first organisms within an evolutionary framework, separating the idea of spontaneous generation from his proposal on a chemical and biochemical emergence of life that incorporated an assemblage of astronomical, chemical and metabolic information and observations in a non-teleological, stepwise evolutionary sequence. Despite the historical significance of his first publication, the small volume is in fact a harbinger to his second book, titled *The Origin of Life*, which appeared in 1936 in Russian and two years later in English (Oparin 1938). This new volume is a refined masterpiece of evolutionary analysis that reflects the process of intellectual, philosophical and scientific maturation of the theory that Oparin had outlined since 1924. As reviewed elsewhere, the new text included a refined discussion of the historical and philosophical issues of the origins of life, leaving behind the crude mechanistic materialism of his first book and incorporating the non-reductive principles of natural law within the framework of a historical narrative (Lazcano 2016).

A detailed analysis of available astronomical and chemical information allowed Oparin (1938) to propose a description of the primitive Earth as a reducing environment, in which the abiotic syntheses of organic compounds eventually led to the appearance of life. He adopted Mendeleyev's proposal on the origin of petroleum and argued that the iron carbides present in the primitive Earth had reacted with water vapor to form hydrocarbons, which, when oxidized, produced alcohols, ketones and aldehydes that, after reacting with ammonia, formed amines, amides and ammonium salts. These compounds became protein precursors, which when dissolved in the primitive hydrosphere formed colloidal systems such as coacervates, from which the first metabolic routes and the first living beings arose (see Lazcano (2016)).

Contrary to what has been claimed (Muller 1966; de Duve 1991), the publication of these first two books did not respond to requests from the Soviet government or the Communist Party. Oparin's heterotrophic theory and his books reflect his interest in developing an evolutionary explanation on the origin and nature of life, and as argued persuasively by Graham (1993), should be seen in the context of the efforts of a society that sought to build a scientific, artistic and cultural culture within the framework of dialectical materialism, but which never imagined the cruelty of the persecutions, censorship and purges that characterized Stalin's government during that period.

Oparin's proposal is not a theory about coacervates or the metabolic origin of life. It is an evolutionary theory about the emergence of living protoplasm where both metabolism and genetic inheritance met. To assume that in his first two books Oparin denied the importance of genetic material is to fail to understand that like many of his contemporaries both in the USSR and in western countries, he advocated a pre-Mendelian vision of genetics. The issue is certainly complicated by Oparin's politically loaded writings and by the partisan attitude he adopted when he actively sided with the positions of Lysenko and the Soviet establishment. But it is equally true that at the time many of his contemporaries doubted the existence of genes, as shown, for instance, by Morgan's (1933) remark in his 1934 Nobel Lecture that: "There is no consensus [...] as to what genes are—whether they are real or purely fictitious".

Led by his leftist sympathies and the belief that science could help in the construction of a more egalitarian society, the distinguished geneticist Hermann J. Muller moved to the USSR in 1993, but soon became disenchanted with the dictatorial attitude of the Bolsheviks and left in 1937. At first, he viewed the heterotrophic theory with considerable sympathy. He wrote:

> [...] the idea, aptly expounded in detail in recent years by Oparin (1938), that there must have been an extended accumulation of ever more complex organic combinations, permitted by the absence of living organisms that would break them down, before genetic material could accidentally arise from them and be suitably provided with the components needed for its own reproduction.

Muller's political disappointment with the Bolsheviks occurred hand in hand with his strong rejection of Oparin's closeness to Lysenko, and eventually gave rise to an intense scientific antagonism echoed in their writings. Muller wrote: "[The suggestion based on Haeckel] that protoplasm had originated first and had the capability of manufacturing not only the genetic material but also its own complex organization", adding that "The Russian Oparin has since the early 1930's espoused this view and has followed the official Communist Party line by giving the specific genetic materials a back seat" (Muller 1966). The intensity of the discussions is a reflection of the Cold War period of anxiety and tension reflected in scientific debates. However, it is also true that Muller was a Mendelian mutationist, not a Darwinian, and had claimed over and over again that life had appeared with the chance emergence of a DNA molecule endowed with the basic properties of living systems. For instance, following the 1959 in vitro synthesis of DNA achieved by Arthur Kornberg and his associates, Muller (1960) stated that "those who define life as I do will admit that the most primitive forms of things that deserve to be called living had already been made in the test tube by A. Kornberg".

Oparin stood his ground, stressing that life cannot be reduced to a single molecule, nor could it be assumed that the haphazard assembly of functional parts of abiotic origin had led to living entities. He wrote:

> [...] it would be wrong to suppose that there first arose proteins, nucleic acids and the other complicated substances found in the protoplasm, which had intramolecular structures which were extremely well and efficiently adapted to the performance of particular biological functions, and that living protoplasm itself arose as the result of a combination of these substances (Oparin 1957).

True to his idea that no substance is alive by itself, Oparin advocated the evolutionary interaction of different kinds of prebiotic compounds, a choice that led him to develop his somewhat naïve original proposals of gels and droplets in 1924. In his 1936/1938 book, he turned toward coacervates as a theoretical and experimental model of compartamentalization. Like many of his contemporaries, Oparin was convinced that the protoplasm could be modeled by the coacervate droplets studied by the Dutch protein chemist Hendrik G. Bungenberg de Jong, who in 1929 described them as colloid systems formed by a liquid phase denser than the environment, where in the absence of a distinct membrane macromolecules of opposite charges form aggregates that can exchange matter with their surroundings.

Like most biologists of his time, Oparin was not aware that cells are not merely membrane-less, protein-rich colloid systems, but are endowed with biochemically active, complex bilayer lipidic membranes. Quite surprisingly, as time went by he maintained his adherence to the Haeckelian concept of "living protoplasm", and until the very of end of his scientific career continued to endorse coacervates as models of protocells (Oparin 1972). Although nowadays the term "protoplasm" is rarely used in biology, coacervates, which are known to exist in eukaryotic cells and may well have formed in the waters of the primitive Earth, are regaining their appeal as models of precellular systems (Aumiller et al. 2016; Smokers et al. 2022).

6.4. Precellular evolution and the emergence of cells

The extraordinary impact of the Miller experiment rested on the elegant demonstration that the abiotic synthesis of organic compounds of biological significance is possible under the environmental conditions proposed by Oparin (Miller 1953). The paper appeared three weeks after Watson and Crick published their model of the DNA double helix, but the study of the origins of life and the nascent field of molecular biology remained separated for almost a decade. The situation changed when Oró (1960) demonstrated that adenine, a nucleobase that plays a central role not only as a component of nucleic acids, but also in the metabolism, could be synthesized from HCN under putative primitive Earth conditions. In other words, the abiotic synthesis of nucleic acid components opened up the possibility of a prebiotic origin of genetic systems. Not surprisingly, during the first 20 years following the 1953 Miller experiment, attempts to understand the origin of life were shaped to a considerable extent first, by the unraveling of the details of DNA replication and protein biosynthesis and, afterward, by the rapid understanding of the molecular biology of cells, including their membranes.

As noted by Strick (2004), in the wake of the launching of the Sputnik and as a result of a complex mixture of social, political, military and scientific interests: "On 29 July 1958, President Eisenhower signed the National Aeronautics and Space Act, creating NASA as the US space agency [...] NASA was formed in 1958, [and it can be seen as] the epitome of Cold War science institutions [...]". The significance of the rapid development of planetary sciences for our understanding of the environmental conditions in which life arose can hardly be overstated. Together with increasingly refined prebiotic simulations, the geochemical analysis of meteorites, cometary dust and asteroid samples material support the contention that prior to the origin of life the primitive Earth was endowed with (a) a vast and diverse collection of organic molecules of biochemical significance; (b) many inorganic and organic catalysts (c) purines and pyrimidines, that is, the potential for template-directed polymerization; and (d) membrane-forming lipidic compounds. This has led time and again to somewhat simplistic models that assume that life is the outcome of the fortuitous association of components preadapted for biochemical functions, including metabolic cycles, the assembly of subcellular structures, and even the genetic code.

The alternative to these overoptimistic proposals is, as Oparin (1957) wrote, to recognize:

> the formation of polymers in the shape of polypeptides and polynucleotides, assemblages having, as yet, no orderly arrangement of amino acid and nucleotide residues adapted to the performance of particular functions. These polymers were, nevertheless, able to form multi-molecular systems, through these were undoubtedly simpler than living protoplasm. It is only by the prolonged evolution of these systems and their interaction with their environment and their natural

selection that there developed the forms of organization characteristic of the living body: metabolism, proteins, nucleic acids and other substances with complicated and 'purposeful' structures which characterize the contemporary living organism.

In conceptual terms, Oparin was right. It is true that the use of condensing agents like cyanimide or hydration/evaporation cycles can lead to the abiotic synthesis of polypeptides and oligonucleotides. However, the refined structural and functional properties of proteins, nucleic acids and other cellular components could not emerge spontaneously, but should be understood as an outcome of an evolutionary process that defies a simple, lineal teleological explanation. For Oparin, of course, such evolution could not take place in single, isolated molecules in classes of molecules, but was the outcome of the evolution of phase-separated chemical assemblages like the coacervate droplets he favored.

The lack of paleontological and molecular records of descent of the first living entities throws us into a swampy terrain full of pitfalls. Extrapolation to prebiotic epochs of extant biochemical components, structures and processes is not guaranteed. For instance, prebiotic pathways are quite different from present metabolic routes, and polypeptides may have resulted not from the dehydration of amino acid solutions, but from a mixture of α-amino- and α-hydroxy-acids (Lin et al. 2021). As summarized elsewhere, backtrack comparisons are impaired by polyphyletic secondary losses, horizontal gene transfer, orthologous displacements and other alternate routes that may have disappeared (Becerra et al. 1997; Lazcano and Miller 1999; Krishnamurthy 2020; Krishnamurthy et al. 2022). Since the emergence of life is marked by the transition from purely chemical reactions to autonomous, self-replicating molecular entities that are capable of evolving by natural selection, some form of inheritance is required.

One of the most likely candidates appears to be RNA or RNA-like molecules. However, for the time being, molecular phylogenies cannot be extrapolated to an evolutionary period prior to the emergence of ribosome-mediated protein synthesis. There is nothing in molecular cladistics that supports the possibility of ancestral life with replicative and metabolic abilities based solely on RNA molecules. Evidence for the so-called RNA world comes from the catalytic versatility of ribozymes and is supported by the biological ubiquity of RNA and ribonucleotides, and not by the extrapolation of the root of phylogenetic trees into the muddy waters of the primitive Earth.

The remarkable coincidence between the monomeric constituents of living organisms and those synthesized in laboratory simulations of the prebiotic environment is certainly too striking to be fortuitous. However, this does not mean that their presence in extant biological systems can be explained as a direct outcome of their prebiotic availability. For instance, the presence of the highly conserved 20 protein α-amino acids encoded by the contemporary genetic apparatus does not mean that they were essential for the origin of life or, much less, that they were the only ones available in the primitive environment. For instance, α-aminobutyric acid, α-amino isobutyric acid, alloisoleucine, norleucine, homoserine and norvaline are found in carbonaceous chondrites, and are major prebiotic products, but are missing in extant proteins (Weber and Miller 1981; Johnson et al. 2008). As reviewed elsewhere, under experimental circumstances norvaline can be synthesized in cells and incorporated in hemoglobin in place of leucine, but only in very specific sites of the protein. This suggests that its presence affects the structure or the folding of hemoglobin and may explain why norvaline is not part of the standard set of biological amino acids (Apostol et al. 1997; Alvarez-Carreño et al. 2013).

In other words, the structure and functional properties of the molecular components of cells, together with the subcellular systems they form and their adaptive complex properties, can only be understood as the historical outcomes of a combination of physical-chemical constraints and natural selection, which reflect the contingent nature of biological evolution. History, however, is not a mere sequence of chance events. Natural selection can overcome contingent effects to an extent, but chemical and biochemical constraints may have a greater significance than is generally recognized. To understand the nature of life, we must recognize both the limits imposed by the laws of physics and chemistry, as well as history's contingency.

The problem of the reconstruction of the events linking the prebiotic soup with the first living beings is equivalent to the issues faced in other historical reconstructions. As noted recently by Bell (2022), "prehistory is more abstract than history: untethered by written documents, it will always be open to guesswork". The same is true for precellular evolution. The issue is complicated not only by the absence of a fossil record of protocellular systems, but also by the lack of an all-embracing generally agreed definition of life. But it is precisely during this poorly understood evolutionary stage when the basic characteristics of living beings arose. It is easy to recognize that the biochemical properties of subcellular systems and their components (such as nucleic acid replication, enzymatic catalysis and membrane-forming properties) are the historical outcome of the evolutionary amplification of the basic physicochemical properties that their simpler molecular predecessors also possessed. Such exquisite functional and structural properties highly adapted to the interaction between different molecules, which Oparin (1957,

1972) defined as "purposiveness", could not have appeared spontaneously in the primitive Earth. The structure and functional properties of the molecular components of cells, together with the subcellular systems they form and their exquisite interaction, can only be understood on the basis of both their physical and chemical characteristics and their adaptive complex interactions, which reflect historical processes of evolution.

These highly refined biological networks are the outcome of the evolution of the system as a whole, and their basic traits must have been established during the poorly defined period that connects the prebiotic phase of syntheses and accumulation of organic compounds with the first biological entities. Life does not reside in one of these components, but depends on their connections and the functional integration between them and the environment. How do we study this ill-defined evolutionary stage? In retrospect, many of Oparin's experiments may appear naïve, but they reflect the scientific knowledge of his time. Quite importantly, compartamentalization processes that may have enhanced concurrent cooperative interaction between different prebiotic components include amphiphilic assemblies, mineral surfaces and liquid–liquid heterogenous phases in aqueous solutions (see Ditzler et al. (2022)), or combinations of them.

The chemical composition of cell plasma membranes, combined with the presence of simple fatty acids in meteorites and the prebiotic synthesis of fatty acids and phospholipids make micelles and liposomes extremely attractive models of protocellular structures (Deamer and Pashley 1989; Luisi et al. 1999; Chen et al. 2004). Lipidic vesicles in which reaction networks are embedded are in fact chemical microreactors in which growth can be coupled to the replication of encapsulated RNA molecules and can exhibit detectable levels of group selection among different compartments. We should not push a model too far, and complementary alternative approaches include, for instance, (a) the critical revaluation of coacervates (Aumiller et al. 2016; Smokers et al. 2022; Köksal et al. 2022) and (b) demonstrations of the spontaneous formation of higher order supramolecular structures by a simple phosphorylation process.

It is of course true that if the origin of life is seen as the evolutionary transition between the non-living and living, then it easy to accept that complex systems of a purely physical and chemical nature played a role in the emergence of life. There are a number of examples of self-organization phenomena that result in supramolecular systems, which may be relevant to the origin of life, and that do not require encoding by genetic polymers. The list includes (a) the formation of micelles, liposomes and lipid vesicles from prebiotic amphiphiles; (b) the self-assembly of nucleic acids (base-bearing polymers); (c) the formation of Fe-S catalytic clusters; (d) the self-assembly of mineral and organic compounds complexes (clays and bases); and (e) autocatalytic synthetic reactions, of which the formose reaction appears to the sole example (Lehn 2002; Orgel 2008; Budin and Szostak 2010; Lazcano 2010).

Regardless of the complexity of the prebiotic environment, life could not have appeared in the absence of a genetic replicating mechanism to guarantee the maintenance, stability and diversification of its basic components under the action of natural selection, that is, the emergence of life was marked by the transition from purely chemical reactions to autonomous, self-replicating molecular entities capable of evolving by natural selection. This requires some form of inheritance, and the most likely candidates appear to be genetic polymers. This does not imply that life can be reduced to a single type of replicating molecule. No molecule is alive by itself, even if it exhibits catalytic and replicative properties. Components co-evolved because they were encoded, that is, did not change in isolation. Other biochemical systems like micelles and liposomes can replicate, and the same is true of prions, but they multiply without transmission of genetic information (Orgel 1992). In other words, the increasingly refined, tight correlation between processes and components is a historical process that may have happened very rapidly in evolutionary times.

Considerable information has been derived from the catalytic and replicative properties of RNA molecules. What is the RNA world? There are many definitions of what it is, including several contradictory ones. We could say that the catalytic, regulatory and structural properties of RNA molecules and ribonucleotides, combined with their ubiquity in cellular processes, suggest that it is an early, perhaps primordial, stage during which RNA molecules played a much more conspicuous role in heredity and metabolism. Even if this possibility is not accepted, the origin and the manifold roles of RNA in extant biology need to be addressed and explained. It is conceivable, for instance, that DNA and RNA appeared simultaneously in the primitive Earth (Bhownik and Krishnamurthy 2019), and that the latter was selected due to pressures related to its functional plasticity.

6.5. Final remarks: does Oparin still matter?

Do we need a new theory on the origins of life? One hundred years have passed since the publication of Oparin's first book on the heterotrophic origin of life, and the cracks and the faults in the original proposal are quite evident. As the history of science demonstrates, no theory remains unchanged. This is especially true in the case of theories that address biological phenomena, such as the cell theory, Mendel's theory of heredity, the endosymbiotic theory and even Darwin's theory of evolution, all of which have been updated. In many ways, this is easy to understand, since the empirical description of biological phenomena depends on historical contexts that change.

Our attempts to understand the emergence of life are limited by the absence of detailed geological evidence, the lack of fossil and molecular evidence of precellular systems emphasized here, and a working, universally accepted definition of life. Perhaps the most important contributions of the heterotrophic theory include Oparin's proposal to set the question of the origin of life within a historical, that is, an evolutionary framework, as well as the detailed scheme of a step-wise, non-teleological process that reinterpreted many isolated facts and observations within a sequence leading to the first organisms. In other words, he defined the origin of life as an evolutionary chemical and biochemical process completely detached from the idea of spontaneous generation. In this sense, as argued by Farley (1977), Oparin's (1938) book is a milestone that may be the most significant work ever published on the emergence of living beings.

It is clear that Oparin's ideas have aged. He proposed a heterotrophic origin of life based on the apparent simplicity of the pathway and the ubiquity of fermentative enzymes, and also suggested that the oldest catabolic route was glycolysis, as it was the only well-known fermentative route at the time. As shown by Clarke and Elden (1980) and Fothergill-Gilmore and Michels (1992), sugar fermentation is the outcome of a patchwork process that recruited enzymes from other processes. Accordingly, a reassessment of the heterotrophic theory would imply the recognition that the first biological systems formed from the compounds present in the primitive environment, and may have depended on cyanamide, thioesters, glycine nitrile and other high energy compounds (see Lazcano and Miller (1999)).

As argued elsewhere, the open character of the theory of chemical evolution and a heterotrophic origin of life has allowed the embodiment of major new discoveries, without destroying its overall structure and premises. This is shown, for instance, by the evolutionary interpretation of the surprising catalytic properties of ribozymes and the different proposals of RNA worlds, as well as by updated descriptions of primitive scenarios that acknowledge the contribution of extraterrestrial organic material. The heterotrophic theory is not about a highly reducing primitive atmosphere or coacervates, but about the gradual (but not necessarily slow), number of evolutionary changes brought about by the emergence of more complex structures and processes. Of course, major gaps can be recognized, but as John D. Bernal wrote in his small volume *The Physical Basis of Life*, our ignorance on the origin of living beings:

> [...] does not mean that we should accept wild hypothesis of the origin of life or of matter, which simply conceal ignorance, but rather that we should attempt almost from the outset to produce careful and logical sequences in which we can hope to demonstrate that certain stages must have preceded certain others, and from these partial sequences gradually built up one coherent history. There are bound to be gaps where this cannot be done, but until the process is attempted these gaps cannot be located, nor can the attempt be made to fill them up… (Bernal 1944).

This is, indeed, a lesson to be kept in mind.

6.6. Acknowledgments

I am deeply indebted to Professor Ernesto Di Mauro for his patience and many conversations on a wide range of issues related to the origin and nature of living systems. It is a pleasure to acknowledge several important exchanges on the scientific biography and contributions of Ernst Haeckel with Prof. Uwe Hobfeld and Dr. Georgy Levit (Friedrich-Schiller Universität Jena). Financial support from UNAM DGAPA-PAPIIT (IN214421) is gratefully acknowledged. I thank Drs. Ricardo Hernández-Morales and José Alberto Campillo-Balderas for their help with the work reported here and the preparation of the manuscript. The results discussed here are based on previously published studies (Lazcano 2016).

6.7. References

Apostol, I., Levine, J., Lippincott J., Leach, J., Hess, E., Glascock, C.B., Weickert, M.J., Blackmore, R. (1997). Incorporation of norvaline at leucine positions in recombinant human hemoglobin expressed in *Escherichia coli*. *J. Biol. Chem.*, 272, 28980–28988.

Aumiller, W.M., Cakmak, F.P., Davis, B.W., Keating, C.D. (2016). RNA-based coacervates as a model for membraneless organelles: Formation, properties, and interfacial liposome assembly. *Langmuir*, 32, 10042–10053.

Bada, J.L. and Lazcano, A. (2003). Prebiotic soup: Revisiting the Miller experiment. *Science*, 300, 745–746.

Becerra, A., Islas, S., Leguina, J.I., Silva, E., Lazcano, A. (1997). Polyphyletic gene losses can bias backtrack characterizations of the cenancestor. *J. Mol. Evol.*, 45, 115–118.

Bell, J. (2022). The world of Stonehenge. *London Review of Books*, 44, 12–15.

Bernal, J.D. (1944). *The Physical Basis of Life*. Routledge and Kegan Paul, London.

Bhowmik, S. and Krishnamurthy, R. (2019). The role of sugar-backbone heterogeneity and chimeras in the simultaneous emergence of RNA and DNA. *Nature Chem.*, 11, 1009–10011.

Budin, I. and Szoztak, J.W. (2010). Expanding roles for diverse physical phenomena during the origin of life. *Annual Review of Biophysics*, 39, 245–263.

Chen, I.A., Roberts, R.W., Szostak, J.W. (2004). The emergence of competition between protocells. *Science*, 305, 1474–1476.

Clarke, P.H. and Elsden, S.R. (1980). The earliest catabolic pathways. *J. Mol. Evol.*, 15, 33–38.

Cleaves, H.J. (2010). The origin of the biologically coded amino acids. *J. Theor. Biol.*, 263, 490–498.

De Duve, C. (1991). *Blueprint of a Cell: The Nature and Origin of Life*. Patterson, Burlington.

Deamer, D.W. and Pashley, R.M. (1989). Amphiphilic components of the Murchison carbonaceous chondrite: Surface properties and membrane formation. *Orig. Life Evol. Biosph.*, 19, 21–38.

Ditzler, M.A., Popovic, M., Zajkowski, T. (2022). From building blocks to cells. In *New Frontiers in Astrobiology*, Thombre, R. and Vaishampayan, P. (eds). Elsevier, New York.

Farley, J. (1977). *The Spontaneous Generation Controversy from Descartes to Oparin*. Johns Hopkins University Press, Baltimore and London.

Fothergill-Gilmore, L.A. and Michels, P.A.M. (1992). Evolution of glycolysis. *Prog. Biophys. Mol. Biol.*, 59, 105–235.

Graham, L.R. (1993). *Science in Russia and the Soviet Union: A Short History*. Cambridge University Press, New York.

Haeckel, E. (1862). *Die Radiolarien (Rhizopoda Radiaria) eine Monographie*. Druck und Verlag, Berlin.

Haeckel, E. (1900). *The Riddle of the Universe*. Watts & Co., London.

Johnson, A.P., Cleaves, H.J., Dworkin, J.P., Glavin, D.P., Lazcano, A., Bada, J.L. (2008). The Miller volcanic spark discharge experiment. *Science*, 322, 404.

Köksal, E.S., Poldsalu, I., Friis, H., Mojzsis, S.J., Bizzarro, M., Gözen, I. (2022). Spontaneous formation of prebiotic compartment colonies on Hadean Earth and Pre-Noachian Mars. *ChemSystemsChem*, 4, e202100040.

Kolchinsky, E. and Levit, G.S. (2019). Reception of Haeckel in pre-revolutionary Russia and his impact on evolutionary theory. *Theory Biosci.*, 138, 73–80.

Krishnamurthy, R. (2020). Systems chemistry in the chemical origins of life: The 18th camel paradigm. *J. Systems Chem.*, 8, 40–62.

Krishnamurthy, R., Goldman, A.D., Liberles, D.A., Rogers, K.L., Tor, Y. (2022). Nucleobases in meteorites to nucleobases in RNA and DNA? *J. Mol. Evol.*, 90, 328–331.

Lazcano, A. (1992). *La Chispa de la Vida*. Editorial Pangea/CONACULTA, Mexico.

Lazcano, A. (2010). Historical development of origins of life. In *Cold Spring Harbor Perspectives in Biology: The Origins of Life*, Deamer, D.W. and Szostak, J. (eds). Cold Spring Harbor Press, Cold Spring Harbor.

Lazcano, A. (2013). A question without answers? *Metascience*, 23, 301–304.

Lazcano, A. (2016). Alexandr I. Oparin and the origin of life: A historical reassessment of the heterotrophic theory. *J. Mol. Evol.*, 83, 214.

Lazcano, A. and Miller, S.L. (1999). On the origin of metabolic pathways. *Journal of Molecular Evolution*, 49, 424–431.

Lehn, J.M. (2002). Toward self-organization and complex matter. *Science*, 295, 2400–2403.

Lin, H., Leman, L.J., Krishnamurthy, R. (2022). One-pot chemical pyro- and tri-phosphorylation of peptides using diamidophosphate in water. *Chem. Sci.*, 13, 13741.

Luisi, P.L., Walde, P., Oberholzer, T. (1999). Lipid vesicles as possible intermediates in the origin of life. *Current Opinion in Colloid and Interface Science*, 4, 33.

Miller, S.L. (1953). A production of amino acids under possible primitive Earth conditions. *Science*, 117, 528.

Morgan, T.H. (1934). The relation of genetics to physiology and medicine. The Nobel Prize Lecture. *The Scientific Monthly*, 41, 5–18.

Muller, H.J. (1960). The Centennial Celebration. Panel One: The origin of life. In *Evolution after Darwin: The University of Chicago Centennial Discussions*, Tax, S. and Callender, C. (eds). The University of Chicago Press, Chicago.

Muller, H.J. (1966). The gene material as the initiator and the organizing basis of life. *Am. Nat.*, 100, 493–502.

Oparin, A.I. (1938). *The Origin of Life*. Macmillan, New York.

Oparin, A.I. (1957). *The Origin of Life on Earth*. Oliver and Boyd, Edinburgh.

Oparin, A.I. (1972). The appearance of life in the Universe. In *Exobiology*, Ponnamperuma, C. (ed.). North-Holland, Amsterdam.

Orgel, L.E. (2008). The implausibility of metabolic cycles in the primitive Earth. *PLoS Biology*, 6, e18.

Oró, J. (1960). Synthesis of adenine from ammonium cyanide. *Biochem. Biophys. Res. Comm.* 2, 407–412.

Rivas, M., Becerra, A., Peretó, J., Bada, J.L., Lazcano, A. (2011). Metalloproteins and the pyrite-based origin of life: A critical assessment. *Orig. Life Evol. Biosph.*, 41, 347–356.

Smokers, I.B.A., van Haren, M.H.I., Lu, T., Spruijt, M.T. (2022). Complex coacervation and compartmentalized conversion of prebiotically relevant metabolites. *ChemSystemsChem*, 4, e202200004.

Strick, J.E. (2004). Creating a cosmic discipline: The crystallization and consolidation of exobiology, 1957–1973. *J. Hist. Biol.*, 37, 131.

Weber, A. and Miller, S.L. (1981). Reasons for the occurrence of the twenty coded protein amino acids. *J. Mol. Evol.*, 17, 273–284.

7

Making Biochemistry-Free (Generalized) Life in a Test Tube

Juan PÉREZ-MERCADER[1,2]

[1] Department of Earth and Planetary Sciences and Origins of Life Initiative Harvard University, Cambridge, Massachusetts, USA
[2] Santa Fe Institute, New Mexico, USA

7.1. Summary

As we are discovering a large number and variety of extrasolar planets, many of which could harbor some form of life, it is time to ask the question: Is biochemistry-based life the only chemical support for life? On Earth, all living systems (i) process information, (ii) metabolize, (iii) self-reproduce and (iv) evolve. These are traditionally associated with the presence of a boundary, metabolism and information-carrying polymers. But, can processes (i)–(iv) take place in a non-biochemical chemical system? We present progress in this area resulting from experiments on system boot-up generated in a one-pot batch reactor during the autonomous chemically controlled non-equilibrium synthesis and self-assembly of functional polymer vesicles from a homogeneous blend of small, non-biochemical molecules. We follow their dynamical evolution that integrates metabolism, growth, reproduction and descent with modification under autonomous chemical control, achieved by implementing a polymerization induced self-assembly (PISA) scenario, which solves the concentration problem and generates an enabling and versatile free-energy gradient. As chemicals ("fuels") are consumed in the polymerization

The First Steps of Life,
coordinated by Ernesto DI MAURO. © ISTE Ltd 2024.

reaction, energy is dissipated and entropy changes result in morphological changes, the expulsion of "spore-like molecules" and joint physicochemical evolution. We monitor the consequences of the copolymer synthesis and the resulting evolution of their molecular self-assembly in "vivo". We find that this transient (or dissipative) self-assembly process leads to vesicles with diameters between 0.5 μm and tens of micrometers. They exhibit several autonomous, emergent, life-like properties, including periodic growth and partial collapse, system self-reproduction, together with homeostasis, competition, chemo-phototaxis and adaptation. We also briefly discuss the extension of the above by combining PISA with click or oscillatory chemistry, which yield higher complexity entities. Together, these results offer insights into completely chemistry-based artificial life, as well as into the origin of proto-cells en route to proto-life and the simplest pre-LUCA living systems. This provides the first material realization of Bernal's "generalized life".

7.2. Introduction and background

Life on planet Earth provides us with the most complex manifestations of matter that we know. It uses chemistry, that is, processes "in which atoms and ions change partners" (Feynman et al. 2005) with their attendant "consequences, both in the structure and composition of matter and in the accompanying energy changes" (Adesnik et al. 2002). Using the subset of carbon chemistry we call biochemistry, living systems on Earth are thermodynamically open chemical systems operating in out of equilibrium conditions. We do not know why life exists on Earth or its origin, and we do not know if it exists in other celestial bodies beyond our 4.567 Ga planet (Morris et al. 2019). The fossil record indicates that it was already present on the Earth about 3.7 Ga ago (Fortey 1999). Phylogenetic analysis demonstrates that the three major domains in which life manifests today, namely, Bacteria, Archaea and Eukaryotes, share a Last Universal Common Ancestor (LUCA). We believe that this ancestor must have been far less complex than what we see today. Living systems range in sizes between micrometers (bacteria) and tens of meters (for aspen). Yet, in spite of their extraordinary diversity and complexity, all living systems display four basic properties or functions (Eigen 1995; Morris et al. 2019; Nurse 2021; Munuzuri and Pérez-Mercader 2022). These functions are as follows: (i) the handling and processing of information (Feynman 1982; Cheetham 2011), (ii) metabolism (Cheetham 2011; Smith and Morowitz 2016), (iii) self-reproduction (Volkenstein 1983) and (iv) evolution, both (a) adaptive and (b) non-adaptive (Lynch 2007). In the environments where they exist, by themselves or in interaction with other living species, all extant living systems universally implement these functions using biochemistry.

In extant life, the "processing of information" consists, on the input, of information (contained in the living system's genome and its molecular machinery; Morris et al. 2019), which is then transformed and output in useful forms with which the system is capable of executing the above functions enabling its existence in its environment. With that information, the living system can harness the energy in its environment, and using simpler molecules found in this environment, it synthesizes the chemical molecules that realize its existence. Eventually, some of these parts are used by the living system to self-reproduce, that is, to make fresh "copies" of itself (its "children"). For many reasons, which include the limited fidelity of molecular copying, self-replication (especially in a complex environment) and molecular and systemic degradation, these may not be perfect "copies". They are viable copies, in the sense that all that is required is that the copies themselves can carry out their own independent lifecycle (these "children", an embodiment of von Virchow's "omnia vita ex vita" dictum (Virchow 1859), can coexist for some time with their parents and join their parents' already existing population as they are born, age, and execute their own lifecycle). Finally, the chemistry of extant living systems can change sufficiently and make populations whose individual living systems evolve following adaptive and/or non-adaptive strategies (Lynch 2007).

How did all this come about? We do not know, and given the historical nature of this question, we may never know with certainty. But most of the current scientific thinking on this revolves around the notion that what we see today is the result of co-evolutionary processes (Knoll 2003). That is, processes involving the coupled evolution of life, its chemistry and environment, that is, the co-evolution of life and its planetary environment. Perhaps there was an evolution from the simple to the complex (and from LUCA to today).

However, because of its specification and definition, LUCA already used extant biochemical molecules and was therefore considerably complex. Hence, the notion that perhaps there were simpler predecessor chemical systems constituting simpler lineages that eventually evolved into more complex systems, and finally closer to LUCA and its epoch. These hypothetical systems are called "protocells" and have become an area of experimental research in recent years (Rasmussen 2008). Presently, and because of the lack of historical evidence, we can only make guesses as to how protocells operated or how they appeared, or if so, how they evolved into lineages that eventually led to LUCA. But we do know that if they existed, they must have been constrained by the laws of physics and chemistry, just like the rest of the planet and the Universe.

But, did protocells exist or could they have existed? And, if they existed, what were they like? What can we learn in the chemistry laboratory today about their de novo synthesis? Is biochemistry necessary or just sufficient to implement the four properties common to all life? Can we put together chemical systems that using small molecules (without biochemistry) display the above properties using chemistry and the processes in which atoms change partners and generate changes in structure and energy to implement them? What would we learn?

If such chemically operated protocells could be built, they would fall into the category and be an example of what Bernal conceptualized (Bernal 1965, 1966) as "generalized, or cosmic, biology". This "generalized life" refers to all chemistry based and operated systems that using simple molecules and physical sources of free-energy such as light, execute (unify, display, express) the four properties or shared functions of biochemistry-based (extant natural) life. That is, generalized life displays/expresses the *same functions* as natural life, but with a *different mechanism*. In this context, protocells, together with chemical evolution, are an important and integral part in the evolutionary journey from non-living to living matter.

A practical framework to tackle these questions is to apply an "outside-in" design strategy for a system (Brooks 1995), in which we start with the general properties of the system and then try to reproduce those properties with a lower level and more detailed structure in the logical or mechanistic hierarchy. Think of the separate functions to be implemented and then find a way to couple them so as to generate a cooperative (emergent) system where "the whole is more that the [simple] sum of its parts" (Bridgman 1936). In theory, a potentially successful strategy to couple and integrate these functions is to identify features shared by the general functions and assemble the system in such a way that the shared attributes are necessary, complementary and accessible to the properties. These common physical (or chemical) features provide a framework over which to link the functions sought after (Lin et al. 2021; Munuzuri and Pérez-Mercader 2022).

Although the above is very conceptual and more easily said than done, an example from the history of technology can illustrate this "outside in" approach. Let us consider artificial powered flight. In nature, we have birds, insects and mammals that fly. They use wings which they flap (Tennekes 1998). This is not what Langley, and seven years later, the Wright brothers did to achieve artificial powered flight. They understood the principles of flight and had a profound understanding of the "border of attack" of a wing, which combined with the effects of a moving air mass generated by a rotating propeller of the right pitch could provide a means of using Bernouilli's principle to lift a plane into flight. They did not use feathers or wing flapping. They used the combination of physical principles that enables flight in natural life, as in birds, bats or insects, but implemented in a totally different (and artificial) fashion with wood, metal, linen, and using the internal combustion engine.

A shared attribute for enabling the handling (processing) of information (property (i) shared by all living systems) and metabolism (property (ii) above) for an open (i.e. out of thermodynamic equilibrium) system to autonomously boot-up is provided by the presence of a free-energy gradient. This requires the information to already be present or generated in the system and enabled, as well as communicated. Given the correct boundary conditions (Munuzuri and Pérez-Mercader 2022), this free-energy gradient implies the existence of an entropy gradient, which, if it has the correct sign, can be used as a means to enable the production of localized order (lowering the entropy), as long as the gradient is maintained (Prigogine 1967) and the available information can be transported for its processing. Assuming a constant temperature, in such a situation the system's entropy could be adjusted by an energy flow from the environment and, depending on the Gibbs free-energy consumption and entropy production rates, eventually the system's disorder would dominate and lead to its self-replication (Murray and Hunt 1993; Volkenstein 2009). In practice, the above translate into the need for the system to be spatially extended within some medium from which it is phase separated for the presence of gradients in chemical potential.

In order to metabolize, the system needs to process molecules it finds in its environment and use them as materials for the synthesis of presumably more complex molecules with which it makes, at least, its own parts and also those that will implement its self-reproduction. As Morowitz (1968) pronounced, and as quoted by Harold (1986, p. 22), "The necessary condition for this is that the system be connected with a source and a sink and the work be associated with a flow of energy from source to sink". For this, the system requires the existence of a free-energy gradient with respect to its external environment in order to perform the synthesis of its own more complex parts (molecules). But the system is an open system in some environment (Prigogine 1967), and with passage of time its entropy will increase and eventually the system will degrade and, possibly, disintegrate and "die".

Of course, extant living systems avoid this fate by generating appropriate copies of themselves through the process of self-reproduction, and the new systems can be called their "children" (Volkenstein 2009). Furthermore, due to imperfections in reproduction or changes occurring within the parents during their lifetimes, these populations will contain individuals which are differently (for the better or the worse) suitable to perform their functions in their environment and can therefore lead to a form of selection. They could, in principle, even compete among themselves or acquire traits from their environment (e.g. some available catalyst) and enter into competitive exclusion (Hardin 1960; Mayr 2001; Martin and Hine 2008; Katla et al. 2023) and a "struggle for life" (Gause 2003).

The required entropy and concentration gradients for information handling and the internal autonomous synthesis by a living system of the substances necessary for its life can, in principle, be provided if the system exists in a volume that partitions (separates) it from the environment in which it exists, and furthermore, if it is capable of entrapping (at least partially) the molecules necessary to sustain the chemical reactions to power it while maintaining its thermodynamically open character.

Can we make/construct/create/design a material system that is somehow capable of autonomously implementing the above at molecular levels? It would need to be a dissipative system in which self-assembly takes place (Halley and Winkler 2008; Riess et al. 2020). Can such material system be implemented with chemistry? Is biochemistry needed?

In the following, we discuss experimental work that, not using biochemistry, demonstrates notable progress and promise in the answer to these questions.

7.3. Laboratory implementation of an artificial autonomous, and self-organized functional system

(I) Booting up a polymer system

Synthetic chemistry's "arithmetic demon", vesicles and "exorcising" the "demon"

The synthesis of chemicals is usually an uphill battle, and even more so in carbon chemistry, as it often requires multistep reactions followed by separation, purification and many other steps. Typically, any of these conversion rates are smaller than 1. Thus, put together, they give rise to an overall conversion rate, which can be exponentially much less than 1. In other words, after several of those steps, the amounts of chemicals that can be produced are very small. This situation is known as the "arithmetic demon" (Ireland 1969; Serratosa 1990; Gilbert and Martin 2016) and ways to combat it include a reduction in the number of synthetic steps, catalysis, "clever" use of the affinity between molecules (as in click chemistry, for example), "stocking up" of the ingredients in very large quantities, or concentrating the reactants into a small volume where the reactions can then take place by avoiding dilution effects.

In life, to deal with the arithmetic demon, nature seems to have opted for a combination of the above, including the use of "small containers" within which the complex reactions required by life take place, making extensive use of liquid–liquid phase separation as it occurs in "colloidal" matter. Of course extremely evolved, extant living systems such as bacteria, archaea or the individual and its cells in the complex system which constitute eukaryotes, have a permeable wall which separates them from their environment and therefore solves the concentration problem by providing a (global) free-energy gradient, while also keeping the system thermodynamically open. These natural membranes are made of phospholipid molecules. These are amphiphilic molecules, that is, block co-polymer molecules (Jones 2013), which, given a solvent (H_2O for life on Earth), combine a solvophilic block back to back with a solvophobic block (Jones 2013; Morris et al. 2019). If they are made of non-lipidic polymer blocks, they are known as "amphiphilic block copolymers" or ABCs. Given a high enough concentration of them in a solution (the critical micelle concentration (CMC)), they will self-organize and adopt equilibrium (Halley and Winkler 2008) self-assembled cooperative configurations into space-partitioning morphologies, such as micelles, tubes, vesicles and a number of ordered structures (Israelachvili 2006). The vesicles in natural life are somewhat permeable compartments (see below) made by membranes formed with phospholipid, lipid and fatty acid related amphiphiles, and containing a liquid "lumen" (when made of non-lipid polymeric amphiphiles, they are also called "polymersomes").

Can we make these molecules from artificial components simpler and easier to synthesize and from simpler components (Rosen 2000) than phospholipids? In extant natural life using simpler parts found in the environment, amphiphiles are synthesized by the living system. This synthesis has to be robust and precise, as the lengths of the solvophilic and solvophobic blocks have to be fairly uniform throughout the membrane. Otherwise, their self-assembly will not lead to uniform membranes, which will contain excessive defects and not be able to confine essential chemical materials at a sufficient level and for long enough times so as to allow for necessary functional activities (Jones 2013).

(II) Polymerization-induced self-assembly

Artificially, polymeric amphiphiles can be synthesized in the laboratory with excellent precision under very mild, oxygen tolerant conditions using the technique of RAFT polymerization (Perrier 2021) applied to a previously available "simple" block. This simpler block is called the macro-Chain Transfer Agent (mCTA). Macro (the "m" in the acronym mCTA) refers to a solvophilic (or solvophobic) tail. The CTA is a small molecule that under the right conditions can transfer monomer molecules from the reaction solution, switching its place with them at the end of the solvophilic (solvophobic) block (the macro part of the mCTA molecule), where they are bound and start a solvophobic block, eventually making it by repetition of the transfer process into a solvophobic (or solvophilic) block with the CTA at the end. Thus, using RAFT polymerization, we could begin with a solution containing an mCTA, one of the ends of which is hydrophilic, and polymerize a monomer on it, also contained in the solution, that leads to a solvophobic block when it is in a polymer chain. As the polymerization reaction continues, this resulting ABC will become more solvophobic and therefore change its collective self-assembly configuration. For example, as the polymerization goes on and molecules of an ABC are produced, their resulting collective morphology can be expected to change from free ABC molecules in solution, giving rise to emergent cooperative associations of the still polymerizing ABCs into micelles, worms and on to vesicles. This (extraordinary) process is implemented in a polymerization-induced self-assembly (PISA) reaction, where starting from a stirred homogeneous and isotropic mixture of chemicals under mild conditions, the chemical reactions do their molecular assembly and construction job "directed" by the information contained in the reacting molecules and their environment, and use chemistry's power to simultaneously generate energy and structural changes in some medium. Furthermore, these changes can be controlled by a number of (not necessarily constant or deterministic) external factors such as light, temperature or reactant flow, including their diffusion, convection and migration (Gong and Pérez-Mercader 2018; Penfold et al. 2019; Pearce and Pérez-Mercader 2020a). An illustration of the chemical and physical pathway is briefly provided in Figure 7.1.

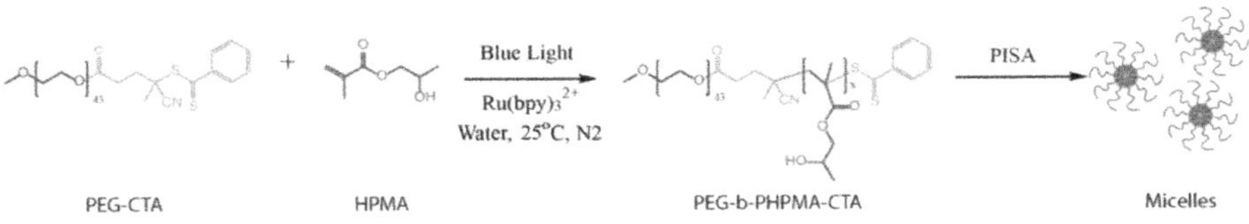

Figure 7.1. *Light mediated polymerization-induced self-assembly. For a color version of this figure, see www.iste.co.uk/dimauro/firststeps.zip*

COMMENT ON FIGURE 7.1.– *Synthesis route for preparation of micelles via PET-RAFT PISA reaction using PEG-CTA and HPMA catalyzed by $Ru(bpy)_3^{2+}$ under blue light irradiation in an oxygen-poor environment. Under the correct environmental conditions, these micelles will evolve into vesicles with a complex behavior that provide an example of non-biochemical, small molecule "generalized" life (Lin et al. 2021).*

Autopoietic synthesis of a functional container (reactor)

As a development of the work in Szymanski and Pérez-Mercader (2016), one such system was presented in Albertsen et al. (2017), where the mCTA is a polyethylene glycol (PEG) polymer with 43 units of EG and has a molecule of CTA containing three sulfur atoms at its end, which can generate the radicals needed for the polymerization. Put in an aqueous solution containing a suitable monomer (Lansalot et al. 2016) such as HPMA (2-hydroxypropyl methacrylate), plus a ruthenium salt to help produce radicals for the polymerization when bathed with blue light, the chemistry in the system generates amphiphiles using RAFT polymerization (Perrier 2021). In the appropriate conditions, these amphiphiles, controlled by the chemistry going on, self-assemble into vesicular structures large enough to be seen in the optical microscope (Szymanski and Pérez-Mercader 2014, 2016). The latter enables the possibility of following the time and morphological evolution of the self-assembled structures under an optical microscope by capturing movies and then quantitatively analyzing and characterizing some of their collective properties (for details, see Albertsen et al. (2017)).

The initial conditions for the reaction are those of a stirred and physically homogeneous blend of chemicals contained in a transparent but otherwise closed vial (batch operation). The first 8 h of the PISA reaction took place in a 10-mL vial at a temperature of 25°C (Figure 7.2(a)). The 5 mL solution went from a stirred transparent mix to a "cloudy" mixture as the various synthesized ABC self-organized and began to form self-assembled structures as a consequence of the ongoing RAFT polymerization process.

This autonomous process generated vesicular structures out of the originally homogeneous and stirred mixture of far simpler unassembled molecules.

After approximately 8 h, a small aliquot of the reacted solution was transferred to a microscope slide where it was illuminated by pulses of blue light lasting 4 s, followed by a 20 ms burst of green light to take the fluorescence image. The resulting image was recorded by a digital camera and a movie was assembled. These images were then analyzed using a combination of standard software techniques and python algorithms.

We clearly see in Figure 7.2(c) that the PISA reaction is generating vesicles with <10 um diameter and smaller objects.

144   The First Steps of Life

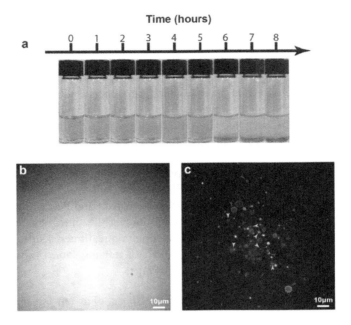

Figure 7.2. *Synthetic autopoiesis: going autonomously from a homogeneous chemical blend to self-reproducing chemical systems. For a color version of this figure, see www.iste.co.uk/dimauro/firststeps.zip*

COMMENT ON FIGURE 7.2.– *(a) The first 8 h of the PISA reaction mixture, left to right; (b) the solution initially after 8 h of blue light illumination in the vial, as seen on the microscope slide; (c) the solution after 3 h of blue light illumination on the microscope. Note the size of the vesicular polymersomes in (c), the sample was stained with 4 µM rhodamine 6 G to record the fluorescence micrographs. Yellow arrowheads in (c) point to monomer droplets within the membrane of the vesicles, and red arrowheads point to polymersomes where the monomer has spread over a larger fraction of the membrane surface (Albertsen et al. 2017).*

7.4. More physics and chemistry working together: phoenix, self-reproduction via spores, population growth and chemotaxis

(I) New emergent phenomena: vesicle growth and collapse

Further study of the time evolution of the vesicles shows another interesting phenomenon. This can be seen in Figure 7.3, which consists of 12 correlative fluorescence micrographs. With the help of the arrowheads printed there, we appreciate that there are vesicles that grow with time (green arrowheads) and then collapse, to then grow again and repeat the cycle. We call this phenomenon "phoenix", because it reminds us of the Greek legend, where the Phoenix creature collapsed into its ashes and revived again. Quantitative study of this behavior can be carried out by measuring and plotting the diameter of the vesicle as a function of time. The results are shown in Figure 4.4.

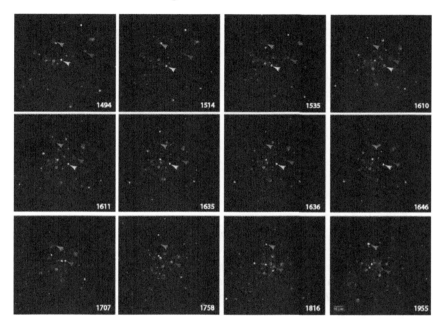

Figure 7.3. *Time lapse of the phoenix behavior of the polymersomes. For a color version of this figure, see www.iste.co.uk/dimauro/firststeps.zip*

COMMENT ON FIGURE 7.3.– *The green arrows highlight growing polymersomes and the yellow and blue arrows point to polymersomes just after their collapse. The scale bar applies to all frames and corresponds to 10 μm. The number in the bottom right of the frame denotes the frame number. The data were collected at five second intervals. The sample is stained with rhodamine 6G (Albertsen et al. 2017).*

We see how the vesicle grows in size until a maximum size is reached and it suddenly collapses. The growth can be almost sigmoidal (as seen in some of the cycles shown in Figure 7.4), but it can also be power law, exponential or logarithmic growth, but always reaching a maximum size (which for a given vesicle decreases its maximum diameter with the number of times the Phoenix cycle is executed), after which the vesicle collapses.

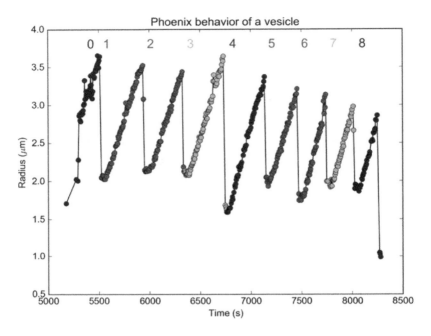

Figure 7.4. *(a) With each successive cycle the maximum size of the polymersome decreases. Different polymersomes were tracked and their area measured using a computer algorithm. The oscillations span thousands of seconds (Albertsen et al. 2017). For a color version of this figure, see www.iste.co.uk/dimauro/firststeps.zip*

Although the vesicle collapse may sound counterintuitive, it is actually the kind of behavior expected on the basis of the Rayleigh–Plesset form of the Navier–Stokes equations for the hydrodynamics of a bubble with its own internal pressure (Plesset and Prosperetti 1977) immersed in a pressure field. In our phoenix vesicles, it leads to a highly nonlinear form of "chemo-cavitation" controlled and brought in by chemistry and the nature of the physics of the materials we are synthesizing. Interestingly, although there are other contributions like changes in surface tension and the presence of defects, the hydrodynamics is mostly powered by PISA, its chemistry and by the osmotic pressure differential building up at the membrane due to different chemical concentrations inside and outside of the vesicles (Lin et al. 2021; Figure 7.5). Interestingly, for a variety of conditions, the Rayleigh–Plesset equation predicts that the "bubble" (vesicle) will undergo successive cycles of growth and collapse.

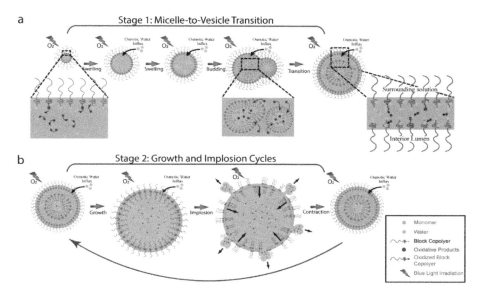

Figure 7.5. *Schematic representation of a "phoenix" cycle. For a color version of this figure, see www.iste.co.uk/dimauro/firststeps.zip*

COMMENT ON FIGURE 7.5.– *(a) Stage 1: under exposure to oxygen and blue light irradiation, supramolecular polymer objects transit from micelles to vesicles through a series of intermediate morphologies due to degradation-induced osmotic water influx. (b) Stage 2: the resulting vesicles undergo size growth-implosion cycles which follow the hydrodynamic evolution predicted by the Rayleigh–Plesset equations for cavitation. After these implosion events, the system undergoes a form of reproduction (not represented here), which involves the release of basic conformational information carrying molecules (Albertsen et al. 2017; Lin et al. 2021).*

(II) Reproduction and increase in population size

The birth, growth and death cycles are seen to be accompanied by an increase in population, probably due to the release during collapse of unreacted mCTA or ABC with the CTA appended at the still active ends of some PHPMA chains (Lin et al. 2021) attached to the PEG. In other words, as they collapse, these vesicles replicate, as indicated by the counting of these structures and shown in Figure 7.6. While at first the structures look like small spherical objects, after some time the number of vesicles that appear in the image (i) suddenly overtakes the droplets, and after approximately 4,000 s into the microscope observation (ii) the number of vesicles grows at a faster rate and with a growth pattern different from the one for the structures in (i). The blue light induces the creation of the phoenix cycles as this vesicle growth rate decreases considerably when the blue light intensity is reduced. For more details, see the Supporting Information in Andersen et al. (2017).

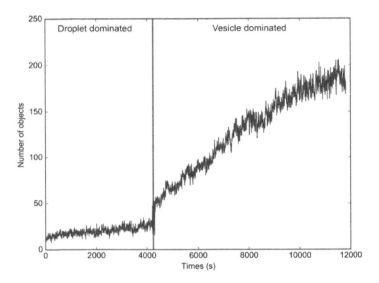

Figure 7.6. *The noisy blue curve shows the change in the type and number of the observed collective structures before ~4,000 s, when the sample was dominated by bright and relatively small filled spherical structures whose number grows at a constant rate, and after ~4,000 s, when the image was dominated by an accelerating and growing number of vesicular structures (Albertsen et al. 2017). For a color version of this figure, see www.iste.co.uk/dimauro/firststeps.zip*

During the predicted hydrodynamic collapse phase of the vesicle, a fraction of the contents of the vesicle is spilled into the surrounding medium. Then it begins to grow again. The subsequently lower osmotic pressure inside the newly reconstituted vesicle can allow for the polymerization reaction in the PISA process to continue, which helped by the inertia of the vesicle's membrane, facilitates a fresh vesicle growth. During the growth and internal dynamical physicochemical evolution of the vesicle, other processes take place inside, which include chemical degradation of the materials in the lumen, as well as polymer sequestration and congestion in the single bilayer membrane, to mention a few important factors. The degradation by reactive oxygen species (ROS) is important because, when it occurs (Lin et al. 2021), it leads to inefficient polymerization due to radical quenching by the ROS, which eventually translates into changes in membrane integrity. The release of inside material into the medium includes incompletely reacted macro-CTA molecules and partially formed amphiphiles. These ("living polymers") will leak out to the vicinity of the imploding vesicle. In the outside medium, they find the materials necessary to complete the amphiphile synthesis and reach a concentration equal or greater than the CMC and thus give rise to new phoenix systems. We can think of the above partially reacted macro-CTA molecules as "seeds" or "spores", since it is from them that the "mother-system" self-reproduces. The cyclic process can continue through a remarkable number of cycles and the parallel with the generic cell division cycle of extant biology represented in Figure 7.7 is clear (see Figure 7.4).

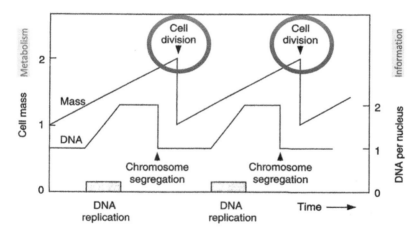

Figure 7.7. *This idealized and abstract depiction of the cell division cycle for living systems shows how in extant life living systems go from one generation to the next in a process that involves the concerted action of metabolism and the change in the information carrying capability of the system. Adapted from Hunt and Murray (1993). For a color version of this figure, see www.iste.co.uk/dimauro/firststeps.zip*

Note that reproduction through "spores" seems to be a less costly process than meiosis and mitosis: less energy and complexity than the others, where reproduction requires a far more complex execution (Svetina 2012). This argues for spores as a more primitive way for reproduction (the hydrodynamics of the spore process are also far simpler than the ones associated with cell division, as discussed above).

Therefore, we see that PISA powered and enabled by chemistry, light and chemical fuels, together with the information contained in the sequence of light pulses and the chemical aliquots in the reacting PISA blend provide a neat, elegant and somewhat simple means to generate, from a completely homogeneous mixture of small molecules, non-biochemical systems with properties (ii) and (iii) at least, common to all natural life. The systems are in addition, self-booting, that is, autopoietic. But, what about properties (i) and (iv)?

(III) Including information and inheritance

In natural life, the control of the chemical system is autopoietic. That is, it is implemented by the internal chemistry of the living system, even though of course it is an open system and therefore is in interaction with its environment, which can affect the system when there are changes in the environment. In extant life, the information is carried in the genome sequence and in the materials that make up the living system: the information transmitted for inheritance uses a polymeric information system. This information is carried in the ordered sequence of the bases in the living system's genome. But this is not the simplest way in which chemical information can be codified, and must have been the result of evolutionary change between the simplest protocells and LUCA. It is highly probable that the simpler forms of protocells did not use a system such as the one we see today.

In the absence of DNA, what could have played the role of information carriers for non-biochemical protocells? What was the reason (and/or vehicle?) why the information had to be passed along from one generation to the next? Presumably the information had to be passed from generation to generation as it would be "very demanding" having to re-invent the wheel in each generation. Or, as Quastler (1964) (implicitly referring to DNA and RNA) put it: "information [in biology] has emerged through the accident of a particular single strand [mutatis mutandis for molecular aliquot combination] becoming the ancestor of the system, that is, through the stability properties of the system descended from that particular single strand". It is a process of "accidental choice remembered". In other words, the necessary information is passed from parent to child, which then inherits it. Both the transmitted information and its vehicle must have been very rudimentary and most probably not in an information storing molecule. Indeed, the information necessary for heredity can be stored in the identity and concentration (aliquots) of the compounds needed for the reaction network to operate and carry out the functions of life (Sagan 1990; Shapiro 2007).

Not surprisingly, our "spores" can play the role of information transmission (heredity) molecules. As previously mentioned, they are incompletely processed ABC molecules whose hydrophobic block is (a) shorter than what it takes to become part of the membrane and (b) carries the CTA attached at its hydrophobic end. These two features make the "spores" in the medium (which still contains enough food) capable of starting new micelles with very similar, but not necessarily identical, properties to the original/mother/parent, which, to begin with, produced the "spore". There is then the potential for variation as, due to their production process, not all "spores" will be identical in length. Therefore, the "spores" can function as carriers of information from one generation to the next.

Furthermore, because the process of "spore" generation is connected with the presence of ROS whose detailed properties are affected by environmental conditions, the "spores" also inherit some non-uniform dependence on the environment, that is, considered as a population of molecules, the "spores" are not all identical due to, among other factors, their polydispersity. The population of "children" generated by the "spores" in one reproductive cycle will then not be uniform, but show some degree of variation. This provides a crude and primitive, albeit practical and effective, form of "descent with modification" (Ridley 2004), already at the level of these small-molecule based "abiotic synthetic protocells".

(IV) Representing life with the methods of theoretical physics

It is interesting to note here that the integration of the four basic properties of life into our non-biochemical system is consistent with the finding that it is possible to represent the properties of livings systems by means of a finite set of equations. As shown in Munuzuri and Pérez-Mercader (2022), it is possible to represent each of the properties (i)–(iv) using a top-level description of the basic mathematical physics and/or chemistry in terms of chemical kinetics. The view in this representation of the living system is that of a system that consists of three theoretical substances representing "food", "cytoplasm" and "waste" configured as a stochastic cubic autocatalytic reaction diffusion system constrained by mass-action kinetics and existing at each of the nodes of an imaginary time-dependent spatial grid. By solving the resulting equations on a computer and analyzing the nature of their numerical solutions we can construct a phase space of their solutions, and the morphologies and phenomenologies they generate. For some ranges of values of the parameters in these equations, we find that there are solutions which incorporate all four properties that characterize life simultaneously. This is important, as it implies that it is not only possible to separately represent the basic properties of life via mathematical functions, which are solutions to some differential equations, but that it is also possible to unify these four properties into a closed and consistent physico-mathematical representation provided by some equations and their solutions, just as natural living systems do. That is, to represent life it is not necessary to have a material realization, let alone biochemical realizations. Thus, it does not come as a surprise to find that we can realize them materially, and chemically in particular, by using a chemistry which eschews biochemistry.

Put together with the above, we can think of the previous example of the chemical synthesis of simple protocells from a homogeneous blend of chemicals as an example of what Bernal (1967) called "generalized life": it integrates the properties of living systems and does not rely on biochemistry.

7.5. Discussion and conclusions

We have seen how starting with a physically homogeneous blend of a few relatively simple organic, but not biochemical, molecules, we can engineer totally autonomous and self-booting chemical systems, which can integrate properties (i), (ii) and (iii) shared by all natural living systems found on Earth. The system is by design completely "biochemistry free". The key enablers are the use of non-equilibrium chemically fueled PISA and the ensuing self-organization and self-assembly of soft matter systems. These systems work by a surprisingly simple combination of physicochemical processes, where physical effects reinforce the order generation capabilities of chemical phenomena: not only does chemistry shuffle atoms around, but with this it generates both structure and collective behaviors of molecules which couple separate regions of physical (real) space: the outside and inside of the active vesicular structure thus erected by chemistry. The systems self-reproduce by some molecular spores, which can implement a rudimentary form of inheritance and "descent with modification" due to the presence of radicals and molecular degradation.

What about properties (iv-a) and (iv-b)? Experiments (Katla et al. 2023) using PISA's need for radicals to operate combined with non-equilibrium phenomena and PISA's dependence on the choice of catalyst and the environment in which it occurs show, not surprisingly, how independently booted-up and then mixed PISA systems, one of which has an advantage provided by the presence of a catalyst, actually compete for resources in a way that closely follows the competitive exclusion principle, one of the basic features of competition in natural life and at the root of Darwin's "struggle for existence". Additionally, by combining PISA with other facile features of small molecular networks of catalysts, we can generate systems that show adaptation at the molecular and system levels (Pearce and Pérez-Mercader 2021). We can envisage their evolution into more complex systems by the co-option of more complex molecules and reaction networks to execute more specialized functions.

In conclusion, artificial life based on actual molecular systems, which combine the exchanges of atomic partners and the generation of structures with the information contained in molecules, their bonds and rearrangements, promises a bright future. The use of PISA to tackle the construction of fully artificial, non-biochemical, chemical mimics of life not only solves the arithmetic demon problem, but also brings with it a powerful connection among out-of-equilibrium phenomena and the properties of living systems. This establishes an interaction of the PISA-generated vesicular system with its environment in a way that enables metabolism, self-replication and the integration with information handling.

These systems provide an embodiment with chemistry and by chemistry of Bernal's "generalized life": they can be seen as a material realization of "same functions, different mechanism" for life as we do not know it.

There is much to be done. The results and techniques described here can find applications in artificial life and astrobiology. In the latter case, these help to understand scenarios where simple lineages may have preceded the appearance of more complex life. They can be used to synthesize new advanced materials or construct self-booting and self-programmed soft robots from a homogeneous mixture under mild conditions. Furthermore, they also offer a solid basis for the implementation of more functionalities and can help advance our understanding of how life emerged in our planet from simple beginnings, and how chemical complexity may have developed here or elsewhere in the Universe.

7.6. Acknowledgments

This work would not have been possible without the discussions and collaborations with many of the members of the author's Harvard group, and especially the ones who have co-authored work cited in this chapter. He would also like to thank Stein Jacobsen, Andrew Knoll, Dimitar Sasselov and Jack Szostak in the Harvard Origins of Life Initiative and the students in the Origins of Life Consortium for many discussions. Discussions with Cyrille Boyer, James Cleaves, Esteban Domingo, James Friar, the late Murray Gell-Mann, Juan Manuel Garcia-Ruiz, Dario Gil, Terry Goldman, Jane Kondev, Herbert Levine, Cormack MacCarthy, Stephen Mann, Susanna Manrubia and Geoffrey West are gratefully acknowledged. The author thanks REPSOL S.A. for their support.

7.7. Appendices: Some additional emergent features in PISA "powered" synthetic biochemistry free protocells

We briefly describe some additional features that emerge in these PISA-generated functional systems. These brief comments are intended as a way to illustrate the possibilities for experimentally constructing "generalized life" protocells and the scenarios that they open for further research and applications. We will briefly mention chemotaxis, the presence of competitive exclusion among artificial "species" of our protocells, the effects that result from introducing an additional feedback loop in the PISA synthesis by combining it with click chemistry, and will close with a brief discussion of the representation of natural life at a very high level (i.e. little detail and abstract), using fundamental principles in theoretical physical chemistry.

7.7.1. Chemotactic behavior

The functional phoenix vesicles also moved to the center of the image and followed a "tumbling" path that is shown in Figure 7.8.

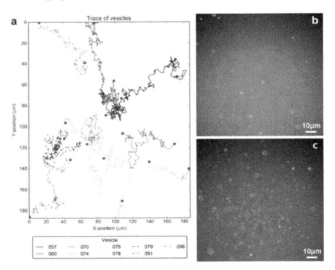

Figure 7.8. *(a) Trace of the movement of a group of vesicles displaying chemotaxis. The collective movement of the vesicle population results in their concentration near the center of the frame where the light was most intense. (b) Micrograph of the solution at the beginning of this experiment and (c) a micrograph at the end of the experiment. The computer trace was generated from a series of micrographs collected during a single experiment (Albertsen et al. 2017). For a color version of this figure, see www.iste.co.uk/dimauro/firststeps.zip*

This motion toward the center of the image could be due to thermal fluctuations or the presence of Marangoni effects (Albertsen et al. 2017) in the vesicles due to the formation of defects in their surfaces as the PISA process continued. The former would be most probable when close to the ABC glass temperature. The latter is most probable when the membrane thickness is not uniform, due to the droplet coalescence processes going inside the vesicle and combined with the chemical reaction being maintained by the blue light irradiation (Gong and Pérez-Mercader 2019, 2020). This is a manifestation of motion due to chemical processes affecting the hydrodynamic equilibrium of the vesicles, that is, a form of chemotaxis. It was also observed in more complex autopoietic systems involving the integration of click chemistry together with PISA (Pearce and Pérez-Mercader 2021b), which we will now turn to.

7.7.2. Adaptive behavior and click-PISA

An important property of living systems is their ability to adapt. In engineering, there are theorems that show that for a system to display adaptive control, there must be two integrated feedback control loops (see refs. 28 and 34 in Pearce and Pérez-Mercader (2021b)) simultaneously present and properly interconnected in the system. This of course requires a more complex system. The PISA system provides a route to increase chemical system complexity, since self-assembly acts as a positive feedback to the RAFT polymerization because phase separation increases the local monomer concentration in the vicinity of the CTA functional groups (see ref. 33 in Pearce and Pérez-Mercader (2021b)), but it starts with a macro CTA already available in the initial reaction vessel. We can ask if we can increase gap in system complexity between the initial PISA reaction and the final self-assembled active vesicular structures by starting with simpler precursor molecules for the macro CTA. This chemical simplification can be achieved (see ACS Central) by integrating two chemical reactions connected by using the radicals in the RAFT based PISA to reduce a pre-catalyst for an orthogonal copper-catalyzed azide-alkyne click (CuAAC) reaction (see ref. 22 in Pearce and Pérez-Mercader (2021b)). The high affinity associated with click actually combines very well with the rest of the RAFT PISA because its high affinity helps avoid the effects of the arithmetic demon before they can actually start. The reduction Cu(II) PMDETA precatalyst by electron abstraction from the RAFT polymerization to activate the CuAAC reaction drives a negative feedback loop. Remarkably, the resulting two-feedback loops system has the same topological as the canonical systems in control theory (see refs. 28 and 34 in Pearce and Pérez-Mercader (2021b)) for describing adaptation in control engineering, and it provides a potential mechanism to chemical adaptivity, where the degree of adaptation in the system to external (light) inputs depends on the relative rates of the associated chemical (click and RAFT) and physical (Self-Assembly) processes. For more details, see Pearce and Pérez-Mercader (2021b).

7.7.3. Competitive exclusion principle and iniferter PISA

An essential feature of natural life is the presence of competition among species or among individuals of a species. The competition among species follows the "competitive exclusion principle" (CEP), which states that any "two species occupying the same niche will compete with each other to the detriment of one of the species, which will thus be excluded" (Hardin 1960; Mayr 2001; Martin and Hine 2008). The CEP is essential for the functioning of Darwin's "struggle for existence". By using an "iniferter" version of the PISA synthesis that we have been discussing (Katla et al. 2023), it is possible to create two (or more) populations of fully functional protocells. These two populations can be chosen so that they only differ in that one of them is endowed with a photocatalyst which confers advantages in reproduction to the population of the species with a photocatalyst, and not to the other, catalyst-free, species. When put in a common environment, the populations compete for common food, and the population with the photocatalyst eventually displaces and eliminates the other one. These of course are biochemistry free systems, yet their competitive exclusion behavior parallels, even in details, the standard competitive exclusion found for paramecia and other species originally found by Gause (2003). The fact that these PISA-generated non-biochemical protocells show competitive exclusion shows that biochemistry (used by natural extant life) is sufficient but not necessary for establishing competitive exclusion and Darwin's struggle for life.

7.7.4. PISA and its control by chemical automata

At some point, we need to integrate the means for strictly chemical control into our system. This requires us to use a manner of control that is chemical at its "root" and capable of carrying out the functions of control associated with computing automata. In other words, we need to implement chemical automata capable of controlling the functioning, and perhaps even the assembly, of a potential "life mimic". Such automata would be "native" chemical automata since once the information is input, the system only has chemistry at its disposal and is fully dependent on chemistry to carry out the required computation.

It turns out that such material native chemical automata at the highest level in the Chomsky hierarchy can be built (Duenas-Diez and Pérez-Mercader 2019a, 2019b, 2020). The use above of the term "material" is important (Minsky 1967; Rich 2008; Linz 2017). It means that by using an actual material chemical realization, we do not have an infinite length tape or an infinite amount of energy at our disposal to operate a potential chemical automaton in the system. The material chemical automata at the highest level in the hierarchy are limited bounded automata (LBAs). They are finite tape Turing machines and are capable of carrying out any computation that can be encoded with context sensitive languages (CSL). For examples of these in bioinformatics of RNA, see Rivas and Eddy (2000). They require two memory stacks (Duenas-Diez and Pérez-Mercader 2019a, 2020; Foulon et al. 2019) and can, for example, be implemented using the chemistry of the Belousov–Zhabotinsky (B-Z) nonlinear oscillatory chemical reaction. The frequency of its redox relaxation oscillations and their amplitudes are effectively stored by the reaction in two different stacks and used to carry out high-level computations (any chemical oscillator can do the same).

7.7.5. Integrating PISA and information control with the Belousov–Zhabotinsky chemical reaction

Thus, in principle, we could exert a good level of chemical control using the B-Z example of a redox oscillator. But there is a bonus: oscillations in chemistry involve the presence of "initiator" and "inhibitor" radicals, and radicals are essential in polymerization. The question is, can the B-Z radicals be used to power a PISA system? The answer is yes, and this was reported in Bastakoti and Pérez-Mercader (2017a), from which Figure 7.9 was extracted. In Figure 7.9(a) we see the redox oscillations and in Figure 7.9(b) we see how the chemical fuel is consumed and powers the PISA reaction. Examples of the vesicles that can be produced are shown in Figures 7.9(c) and (d).

Additional work (Bastakoti and Pérez-Mercader 2017b) showed how the B-Z reaction can not only power PISA, but also induce deformation and blebbing of vesicles (Figure 7.10). The latter can be interpreted as the effects of osmotic pressure on a vesicle which is "drinking" water or (not necessarily alternatively) whose membrane has coupled to the internal B-Z reaction in such a way that energy from the reaction is dissipated on the membrane (Seifert et al. 1991).

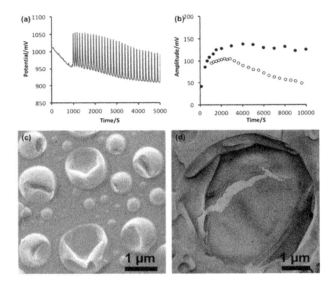

Figure 7.9. *(a) Oscillation profile of the B-Z reaction during amphiphile polymerization between a PEG-CTA and butyl acrylate monomers; (b) amplitude of redox oscillations of the pure B-Z solution (•) and B-Z in polymerization (o). (c and d) SEM image of polymer vesicles after 120 min of polymerization. Adapted from Bastakoti and Pérez-Mercader (2017a). For a color version of this figure, see www.iste.co.uk/dimauro/firststeps.zip*

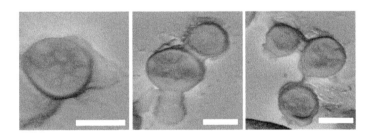

Figure 7.10. *SEM (scanning electron microscopy) images of vesicles showing their appearance and growth of blebs after 80 min of polymerization. The scale bar is 5 μm. These budding and blebbing events result from the reduction of the total interfacial energy due to the B-Z reaction, combined with the gain in interfacial energy due to bud formation (Bastakoti and Pérez-Mercader 2017b)*

7.8. References

Adesnik, M., Fisher, B.A., Larive, C., Li, C., Matayjaszewski, K., Spector, A., White, M.A., Wnek, G. (eds) (2002). *McGraw-Hill Concise Encyclopedia of Chemistry*, 9th ed. McGraw-Hill, New York.

Albertsen, A.N., Szymański, J.K., Pérez-Mercader, J. (2017). Emergent properties of giant vesicles formed by a polymerization-induced self-assembly (PISA) reaction. *Scientific Reports*, 7, 41534.

Bastakoti, B.P. and Pérez-Mercader, J. (2017a). Facile one pot synthesis of functional giant polymeric vesicles controlled by oscillatory chemistry. *Angew. Chemie Int. Ed.*, 56, 12086–12091.

Bastakoti, B.P. and Pérez-Mercader, J. (2017b). Autonomous ex novo chemical assembly with blebbing and division of functional polymer vesicles from a "homogeneous mixture". *Adv. Mater.*, 29, 1704368.

Bernal, J.D. (1965). Molecular structure, biochemical function, and evolution. In *Theoretical and Mathematical Biology*, Waterman, T.H. and Morowitz, H. (eds). Random House, New York.

Bernal, J.D. (1967). *The Origin of Life*. The World Publishing Company, Cleveland, OH.

Bridgman, P.W. (1936). *The Nature of Physical Theory*. Dover Publications, New York.

Brooks, F.P. (1995) *The Mythical Man Month: Essays on Software Engineering*. Addison-Wesley, Reading.

Cheetham, N.W.H. (2011). *Introducing Biological Energetics. How Energy and Information Control the Living World*. Oxford University Press, Oxford.

Cheng, G. and Pérez-Mercader, J. (2018). Polymerization-induced self-assembly for artificial biology: Opportunities and challenges. *Macromolecular Rapid Communications*, 40(2), 1800513.

Cheng, G. and Pérez-Mercader, J. (2019). Engineering programmable synthetic vesicles with permeability regulated by a single molecular bridge. *Chemistry of Materials*, 31(15), 5691–5698.

Cheng, G. and Pérez-Mercader, J. (2020). Dissipative self-assembly of dynamic multicompartmentalized microsystems with light-responsive behaviors. *Chem*, 6, 1160–1171.

Dueñas-Diez, M. and Pérez-Mercader, J. (2019a). How chemistry computes: Language recognition by non-biochemical chemical automata. From finite automata to turing machines. *iScience*, 19, 514–526.

Dueñas-Diez, M. and Pérez-Mercader, J. (2019b). Native chemical automata and the thermodynamic interpretation of their experimental accept/reject responses. In *The Energetics of Computing in Life and Machines*, Wolpert, D.H., Kempes, C., Grochow, J.A., Stadler P.F. (eds). SFI Press, Santa Fe, NM.

Dueñas-Diez, M. and Pérez-Mercader, J. (2020). In-vitro reconfigurability of native chemical automata, the inclusiveness of their hierarchy and their thermodynamics. *Scientific Reports*, 10, 6814.

Eigen, M. (1995). What will endure of 20th century biology? In *What is Life? The Next Fifty Years: Speculations on the Future of Biology*, Murphy, M.P. and O'Neill, L.A.J. (eds). Cambridge University Press, Cambridge.

Feynman, R.P. (1982). Simulating physics with computers. *Int. J. Theor. Phys.*, 1, 467–488.

Feynman, R.P., Leighton, R.B., Sands, M. (2005). *The Feynman Lectures on Physics, The Definitive Edition, vol. I*. Pearson/Addison-Wesley, Reading.

Fortey, R. (1999). *Life. A Natural History of the First Four Billion Years of Life on Earth*. Vintage Books, New York.

Foulon, B.L., Liu, Y., Rosenstein, J.K., Rubenstein, B.M. (2019). A language for molecular computation. *Chem*, 5, 3017–3019.

Gause, G.F. (2003). *The Struggle for Existence*, 1st ed. Dover Publications, Mineola, NY.

Gilbert, J.C. and Martin, S.F. (2016). *Experimental Organic Chemistry: A Miniscale & Microscale Approach*, 6th ed. Cengage Learning, Boston, MA.

Halley, J.D. and Winkler, D.A. (2008). Consistent concepts of self-organization and self-assembly. *Complexity*, 14, 10–17.

Hardin, G. (1960). The competitive exclusion principle. *Science*, 131, 1292–1297.

Harold, F.M. (1986). *The Vital Force: A Study of Bioenergetics*. W. H. Freeman and Company, New York.

Ireland, R.E. (1969). *Organic Syntheses*. Prentice-Hall, Inc., Englewood Cliffs, NJ.

Israelachvili, J. (2011). *Intermolecular and Surface Forces*, 3rd ed. Academic Press-Elsevier, Amsterdam.

Jones, R.A.L. (2013). *Soft-Condensed Matter*. Oxford University Press, New York.

Katla, S.K., Lin, C., Pérez-Mercader, J. (2023). Competitive exclusion principle among synthetic non-biochemical protocells. *Cell Reports Physical Science*, 4, 101359.

Knoll, M. (2003). *Life on a Young Planet: The First Three Billion Years of Evolution on Earth*. Princeton University Press, Princeton, NJ.

Lansalot, M., Rieger, J., D'Agosto, F. (2016). Polymerization-induced self-assembly: The formation of controlled radical polymerization to the formation of self-stabilized polymer particles of various morphologies. In *Macromolecular Self-Assembly*, 1st ed, Billon, L. and Borisov, O. (eds). John Wiley & Sons, Inc, Hoboken, NJ.

Lin, C., Katla, S.K., Pérez-Mercader, J. (2021). Photochemically induced cyclic morphological dynamics via degradation of autonomously produced, self-assembled polymer vesicles. *Commun. Chem.*, 4, 25.

Linz, P. (2017). *An Introduction to Formal Languages and Automata*, 5th ed. Jones & Bartlett Learning, Burlington, MA.

Lynch, M. (2007). *The Origins of Genome Architecture*. Sinauer Associates Inc., Sunderland, MA.

Martin, E. and Hine, R. (2008). *A Dictionary of Biology*, 6th ed. Oxford University Press, Oxford.

Mayr, E. (2001). *What Evolution I*. Basic Books, New York.

Minsky, M.L. (1967). *Computation: Finite and Infinite Machines*. Prentice Hall, Englewood Cliffs, NJ.

Morowitz, H. (1968). *Energy Flow in Biology. Biological Organization as a Problem in Thermal Physics*. Academic Press, New York and London.

Morris, J., Hartl, D., Knoll, A., Lue, R., Michael, M., Berry, A., Biewener, A., Farrell, B., Holbrook, N.M., Heitz, J. et al. (2019). *How Life Works*, 3rd ed. W. H. Freeman & Company, New York.

Munuzuri, A.P. and Pérez-Mercader, J. (2022). Unified representation of life's basic properties by a 3-species stochastic cubic autocatalytic reaction-diffusion system of equations. *Phys. Life Revs.*, 41, 64–83.

Murray, A. and Hunt, T. (1993). *The Cell Cycle. An Introduction*. Oxford University Press, New York.

Nurse, P. (2021). *What Is Life?: Five Great Ideas in Biology*, 1st American ed. W. W. Norton and Co., New York.

Pearce, S. and Pérez-Mercader, J. (2021a). PISA: Construction of self-organized and self-assembled functional vesicular structures. *Polymer Chem.*, 12, 29–49.

Pearce, S. and Pérez-Mercader, J. (2021b). Chemoadaptive polymeric assemblies by integrated chemical feedback in self-assembled synthetic protocells. *ACS Cent. Sci.*, 7, 1543–1550.

Penfold, N.J.W., Yeow, J., Boyer, C., Armes, S.P. (2019). Emerging trends in polymerization-induced self-assembly. *ACS Macro Lett.*, 8, 1029–1054.

Perrier, S. (2017). 50th anniversary perspective: RAFT polymerization – A user guide. *Macromolecules*, 50, 7433–7447.

Plesset, M.S. and Prosperetti, A. (1977). Bubble dynamics and cavitation. *Ann. Rev. Fluid Mechanics*, 9(1), 145–185.

Prigogine, I. (1967). *Introduction to Thermodynamics of Irreversible Processes*, 3rd ed. John Wiley and Sons, New York.

Quastler, H. (1964). *The Emergence of Biological Information*. Yale University Press, New Haven, CT, and London.

Rasmussen, S., Bedau, M.A., Chen, L., Deamer, D., Krakauer, D.C., Packard, N.H., Stadler, P.F. (eds) (2008). *Protocells: Bridging Nonliving and Living Matter*. MIT Press, Cambridge, MA.

Rich, E. (2008). *Automata, Computability, and Complexity. Theory and Applications*. Pearson Prentice Hall, Upper Saddle River, NJ.

Ridley, M. (2004). *Evolution*, 3rd ed. Blackwell Publishing Company, Malden, MA.

Riess, B., Grötsch, R.K., Boekhoven, J. (2020). The design of dissipative molecular assemblies driven by chemical reaction cycles. *Chem*, 6, 552–578.

Rivas, E. and Eddy, S. (2000). The language of RNA: A formal grammar that includes pseudoknots. *Bioinformatics*, 16, 334–340.

Rosen, R. (2000). *Essays on Life Itself*. Columbia University Press, New York.

Sagan, C. (1990). Life. In *The New Encyclopedia Britannica*, 15th ed., vol. 22. Encyclopedia Britannica, Inc., Chicago, IL.

Seifert, U., Berndl, K., Lipowsky, R. (1991). Shape transformations of vesicles: Phase diagrams for spontaneous-curvature and bilayer-coupling models. *Phys. Rev. A*, 44, 1182–1202.

Serratosa, F. (1990). *Organic Chemistry in Action: The Design of Organic Synthesis*. Elsevier, Amsterdam.

Shapiro, R. (2007). A simpler origin for life. *Sci. Am.*, 296, 46–53.

Smith, E. and Morowitz, H. (2016). *The Origin and Nature of Life on Earth: The Emergence of the Fourth Geosphere*. Cambridge University Press, Cambridge.

Svetina, S. (2012). On the vesicular origin of the cell cycle. In *Genesis: In the Beginning*, Seckbach, J. (ed.). Springer, New York.

Szymanski, J.K. and Pérez-Mercader, J. (2014). Straightforward synthesis route to polymersomes with simple molecules as precursors. *Langmuir*, 30, 11267–11271.

Szymanski, J.K. and Pérez-Mercader, J. (2016). Direct optical observations of vesicular self-assembly in large-scale polymeric structures during photocontrolled biphasic polymerization. *Polymer Chemistry*, 7, 7211–7215.

Tennekes, H. (1998). *The Simple Science of Flight*. MIT Press, Cambridge, MA.

Virchow, R.L.C. (1859). *Cellular Pathology*. John Churchill, London.

Volkenstein, M.V. (1983). *General Biophysics*, vol. 1. Academic Press, New York.

Volkenstein, M.V. (2009). *Entropy and Information*. Birkhauser, Boston, MA.

8

Hydrothermalism for the Chemical Evolution Toward the Simplest Life-Like System on the Hadean Earth

Kunio KAWAMURA
Department of Human Environmental Studies,
Hiroshima Shudo University, Japan

8.1. Introduction

8.1.1. Realistic life-like systems on the Hadean Earth

The discovery of functional RNA molecules led to the RNA world hypothesis (Cech et al. 1981; Gilbert 1986). For constructing the most primitive life-like system, functional RNA molecules played important roles in the emergence of primitive life-like system on the Earth (Joyce 2002; Orgel 2004). The formation of an assignment mechanism between molecules preserving both biological information and function in a life-like system is an essential step in the emergence of a life-like system (Figure 8.1) (Eigen 1971; Engen and Shuster 1978; Nemoto and Husimi 1995; Nemoto et al. 1997). In modern organisms, information flow occurs primarily from DNA to protein sequences via the assignment mechanism (Crick 1970). Other molecules are produced predominantly by protein enzymes via indirect assignment, where each enzyme controls a corresponding single reaction (Kawamura, 2016). As RNA molecules preserve both information and enzymatic functions in modern organisms, a chemical network consisting of a set of RNA

The First Steps of Life,
coordinated by Ernesto DI MAURO. © ISTE Ltd 2024.

molecules was exposed during Darwinian evolution, where functional RNA molecules were replicated by RNA-dependent RNA polymerase ribozyme (RP-ribozyme) (Johnston et al. 2001; Zaher and Unrau 2007; Horning and Joyce 2016; Tjhung et al. 2020). RP-ribozyme is a key molecule for constructing such primitive life-like systems; it amplifies different functional RNA molecules to be selected by natural selection pressure. Thus, several studies have been conducted to create artificial RP-ribozymes using an in vitro selection technique (Ellington and Szostak 1990; Tuerk and Gold 1990). Other molecules resembling RNA could also have been involved in the chemical evolution of RNA toward a life-like system (Bolli et al. 1997). Investigations involving the chemical simulation of protein-like molecules indicate that amino acids, peptides, and peptide-like molecules can be formed readily in prebiotic environments compared with RNA molecules (Miller 1953; Fox and Harada 1958); furthermore, they are more stable than RNA molecules in simulated primitive Earth environments (Kawamura 2004). However, peptides or peptide-like molecules cannot preserve biological information (Lee et al. 1996; Ikehara 2009).

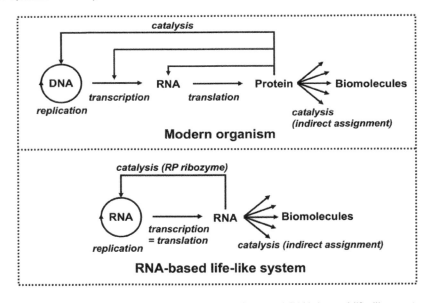

Figure 8.1. *Information flow in modern organism and RNA-based life-like system*

Nevertheless, the RNA world hypothesis should be compatible with the Hadean Earth environment, which involves extreme conditions at high temperatures and pressures (Kawamura 2004, 2012a, 2017, 2019). According to our investigations, strong evidence suggesting that the RNA-based life-like system is incapable of adapting to hydrothermal conditions is currently lacking (Kawamura 2004, 2010, 2019). Although the environments on the Hadean Earth are not very clear (Kasting 1993; Sleep 2010; Maruyama et al. 2013), recent investigations have provided important information. For instance, although only 70% of the Earth's surface is currently covered by water, the surface could have been completely covered by water on the Hadean Earth (Bada and Korenaga 2018; Korenaga 2021). The presence of zircon suggests the presence of water on Hadean Earth, which can be traced back to ca. 4.3 Ga (Mojzsis et al. 2001; Harrison 2009; Sleep 2010). This implies that the surface temperature of the Earth decreased at the very beginning of the Earth, following the accumulation of water to form oceans or freshwater ponds (Santosh et al. 2017). Thus, hydrothermal environments could have played an important role in chemical evolution at the very beginning of the Hadean Earth.

Generally, mild environments are considered for investigations regarding the chemical evolution of RNA (Kawamura 2012). Molecular biological investigations regarding the RNA world hypothesis have been conducted exclusively under mild conditions. This could be attributed to the following two reasons. First, a prejudice that such biopolymers cannot form life-like systems under extreme conditions may have existed. Second, research tools to investigate the possibility of an RNA world under extreme conditions may have been insufficient. With the continuous development of research tools for chemical reactions under hydrothermal conditions, different types of research tools to examine the characteristics and behaviors of RNA and other molecules currently exist (Kawamura 2017). These tools can be applied to determine whether ancient environments on Earth were compatible with the RNA world hypothesis.

8.1.2. *Water in universe*

Hydrothermal conditions are a range of conditions of liquid water; they normally refer to high temperatures and pressures beyond the boiling point of water at the atmosphere and below the critical point of water (647.10 K, 22.064 MPa) (Figure 8.2). This wide range of conditions is believed to be a possible environment during the chemical evolution on Earth and in celestial bodies. Water has existed on the Earth surface since ancient times and is found in the solar bodies, such as Mars, Europa and Enceladus (Hsu et al. 2015). This suggests that liquid water is widely present in the celestial bodies of the universe. In addition, among all the materials present in the body of cell-type organisms, water has the highest percentage. An important role of water in cells is that it serves as a chemical medium for sustaining biochemical reactions and functions. Water is inevitable for life on Earth; thus, it should be a suitable medium for the emergence of life.

In addition, some evidence suggests that water is inevitable for the emergence of life from the viewpoints of chemistry, Earth science and astronomy. First, hydrogen and oxygen atoms are the primary and third most abundant atoms in the universe, respectively. This provides an important background for the presence of water in the universe. Second, from the viewpoint of chemistry, liquid materials are essential for the construction of life-like systems. Essentially, gas or solid materials are not preferable for the formation of life-like systems. This is because in the gas phase, biochemically important weak interactions are ineffective, whereas in the solid phase, the reaction of the molecules may not proceed smoothly.

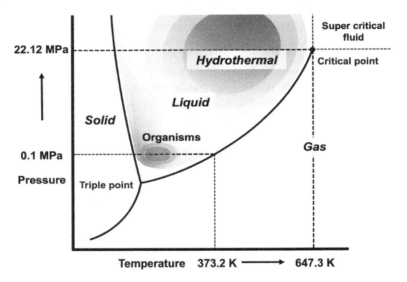

Figure 8.2. *Phase diagram of water. For a color version of this figure, see www.iste.co.uk/dimauro/firststeps.zip*

Next, compared with other liquid materials, water is reasonably suitable as a liquid material for the emergence of life-like systems as highlighted in a previous study (Brack 1993). The organisms that exist currently can survive and grow on Earth under a wide range of temperatures and pressures where water exists as a liquid. Another possibility is the presence of organisms in the remaining range of temperatures and pressures, particularly hydrothermal conditions (Figure 8.2), which have not yet been evaluated. Herein, the critical points and range between the melting and boiling points in the atmosphere are summarized for water and other liquids (Figure 8.3, left and right). Notably, the temperature range between the melting and boiling points of water at atmospheric pressure is relatively high and wide compared with that of NH_3, CH_4, H_2S and H_2Se. The relatively higher melting and boiling points of water are owing to the strength of the hydrogen bonding between the lone pairs of oxygen. A higher temperature range is preferable from the viewpoint of the reaction rate for chemical evolution. In general, chemical reaction rates at low temperatures in liquid materials are much lower than those in water. Currently, some hyperthermophic organisms are known to survive beyond the boiling point of water (100°C) at 0.1 MPa (Kashefi and Lovley 2003; Takai et al. 2008). However, the suitable temperatures for the present organisms are fairly close to the triple point of water, compared with the hydrothermal range. Naturally, different liquid materials, such as CH_4 and NH_3, could have possibly played a similar role of chemical medium for the emergence of life; however, their triple and critical points are very low compared with those of water. Simulation experiments in CH_4 and NH_3 are not popular; therefore, comparative chemical evolution experiments in such liquids are important.

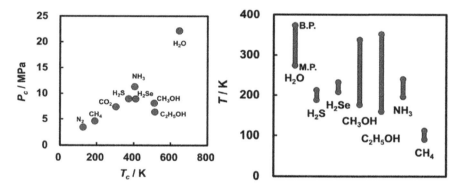

Figure 8.3. *Chemical properties regarding materials to exist as liquid. Left, Pc: critical pressure, Tc: critical temperature; right, M.P.: melting point, B.P.: boiling point at atmospheric pressure. For a color version of this figure, see www.iste.co.uk/ dimauro/firststeps.zip*

The habitable zone is dependent on the distance between the planet and its main star as this zone is estimated based on the distance that enables water to exist as liquid (Kasting 1997). Recent studies have indicated that small satellites located far from the sun possess liquid water (Hsu et al. 2015). In such celestial bodies, the conditions that enable water to exist in the liquid state should be satisfied, even if they are located very far from the sun. In addition, the amount of water and the habitable zone are other important factors. In general, the presence of larger amounts of materials could provide a greater chance for chemical evolution.

Materials, including water and others on Earth, would be sufficient to provide a notable probability for chemical evolution within a reasonable time period. Simultaneously, other conditions in the aqueous phases, such as salt concentration and pH, must also be considered.

8.1.3. Two-gene hypothesis, minerals and high temperature

The RNA-based life-like system still has drawbacks regarding the compatibility of RNA world hypothesis with the environments on the Hadean Earth (Kawamura 2004, 2012a, 2017, 2019). First, although RP-ribozyme may have been important, the initial RNA chemical network is not identified. The interaction between genetics and metabolism is an important issue for the emergence of life-like system (Saladino et al. 2012). Second, the formation of a compartment or isolation mechanism as an integrated system consisting of different functions of RNA molecules has not been clearly shown. Third, the possibility of using RNA-based life-like systems under hydrothermal conditions at a high temperature, corresponding to tracing back the trajectory of temperature and the chemical evolution of RNA, has not been clarified.

8.1.3.1. Two-gene hypothesis

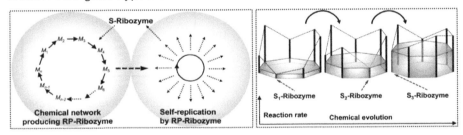

Figure 8.4. *Mechanistic diagrams for the two-gene hypothesis. Left, connection between RP-ribozyme and S-ribozyme; right, the formation of S-ribozyme gives feeds back to the overall rate of the replication. For a color version of this figure, see www.iste.co.uk/dimauro/firststeps.zip*

First, based on the two-gene hypothesis, the linkage between RP-ribozyme and another ribozyme (S-ribozyme) was assumed as the initial step for the formation of a chemical network between genetic information and metabolism (Figure 8.4), as summarized in previous studies (Kawamura 2016, 2019). Ideally, RP-ribozyme can replicate different sequences of RNA molecules. In addition, the reactant monomers for the formation of the RP-ribozyme are formed in a prebiotic metabolic network constructed by chemical evolution, which affects the performance of the ribozyme (Figure 8.4, left). If the slowest step is accelerated to a rate higher than that of the second slowest step via the formation of a ribozyme (S_1-ribozyme), the enhancement of the circular network to form RNA molecules by the S_1-ribozyme feeds back to the overall reaction rate of the replication system (Figure 8.4, right). S_1-ribozyme was postulated to form randomly through the replication of RNA molecules. Thus, S_1-ribozyme was necessary as an initializing ribozyme that links the chemical networks between the metabolic network and replication mechanism by RP-ribozyme.

8.1.3.2. RNA-based life-like system on mineral

Second, the mineral surface is well known to play an important role in the chemical evolution of RNA (Costanzo et al. 2007; Cleaves et al. 2014). In addition, compartment seems to be an essential mechanism for providing stable conditions for a chemical network that behaves as an integrated unit (Szostak et al. 2016; Joyce and Szostak 2018). In fact, the cell type compartment is very complicated; therefore, a simpler mechanism for the compartment should have formed prior to the cell type compartment (Figure 8.5). Thus, the mineral surface is postulated to provide a chemical environment for integrating functional RNA molecules on its surface (Kawamura 2012b; Kawamura et al. 2022a). This is regarded as a mechanism corresponding to the compartment prior to the cell type compartment.

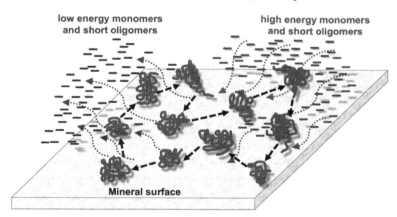

Figure 8.5. *Mineral surface provides a chemical environment for integrating functional RNA molecules on its surface. For a color version of this figure, see www.iste.co.uk/dimauro/firststeps.zip*

8.1.3.3. *The temperature limit for RNA-based life-like system*

Third, kinetic data regarding the prebiotic degradation, formation of RNA molecules and ribozyme reactions were accumulated. In addition, the weak interactions of biomolecules at temperatures above 100°C were measured. Although the upper temperature limit of life was not well identified (Brock 1985), these investigations revealed that the relationship between the decrease in temperature on the Hadean Earth and chemical evolution is important for estimating how and when RNA-based life-like systems were constructed (Kawamura 2010, 2012b, 2019).

In this chapter, the importance of hydrothermal environments, development of research tools and possible RNA-based life-like systems under hydrothermal conditions are discussed. Furthermore, a possible scenario for an RNA-based life-like system under hydrothermal environments is described.

8.2. Hydrothermal environment for the chemical evolution of biomolecules

8.2.1. *As an energy source*

From a thermodynamic viewpoint, life-like systems on Earth can be regarded as dissipative structures. Essentially, life-like systems continuously receive energy and material supplies, and discharge waste. High-temperature conditions can lead to the production of high-energy chemicals. Essentially, the gap between high and low temperatures can be used to produce high-energy products in hydrothermal systems (Figure 8.6). A transitional product formed under hydrothermal conditions, which is stable at high temperatures and pressures, is released and accumulates in the environment at low temperatures. Such chemicals are quickly and slowly degraded into stable compounds after a certain period. These chemicals are regarded as resources located in the upper energy stream for sequential chemical processes to form higher biomolecules. Some evidence suggests the presence of hydrothermal systems on primitive Earth (Martin et al. 2008; Westall et al. 2018). Although the temperature of the primitive ocean is not clearly known (Sleep 2010), the temperature of the hydrothermal system can be reasonably assumed to have been higher than that of the surrounding ocean or freshwater (Westall et al. 2018). Thus, hydrothermal systems are regarded as suitable chemical environments for providing energy sources, as well as other high-energy sources, such as electric discharges, UV and radioactivity.

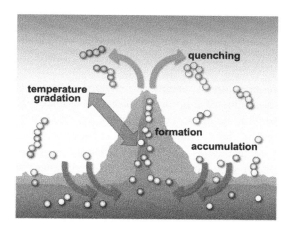

Figure 8.6. *Role of hydrothermal vent system in deep ocean on primitive Earth. For a color version of this figure, see www.iste.co.uk/dimauro/firststeps.zip*

8.2.2. *Temperature and pressure*

Thermodynamic data on the characteristics of water at high temperatures and pressures have been investigated in relation to fundamental chemical studies of the solubilities of minerals and metals, and the dissociation constants of acids (Martell and Smith 1974). Although these investigations were performed using conventional potentiometry, information on the chemical behavior of biomolecules, such as functional peptides and nucleic acids, is scarce. Thus, the expedition to wider temperature and pressure ranges is valuable because the chemical environment of hydrothermal systems is unique compared with low-temperature liquid water. For instance, the dielectric constant of water decreases considerably with increasing temperature (Uematsu and Frank 1980) and becomes similar to that of organic solvents. In addition, the ionic product of water and the strength of hydrogen bonding and hydrophobic interactions vary with increasing temperature (Marshall and Franck 1981).

The reaction rate also increases with increasing temperature. For instance, degradation by hydrolysis of organic molecules occurs readily at higher temperatures. The increase in such reaction rates with temperature is important because the reaction rates normally increase by two to three times with an increase of 10°C. This remarkably enhances the reaction rates; for instance, the rate increases by 2^{10}–3^{10} times (approximately 1,000–60,000-fold) by increasing the reaction temperature from 0 to 100°C. The effect of increasing the temperature is dependent on the enthalpy and entropy changes, thus affecting the formation and degradation of organic molecules. In particular, both peptides and nucleic acids undergo dehydration in the aqueous phase in modern organisms via a high-energy activation mechanism.

In addition, pressure affects chemical reactions (Maurel et al. 2020). The influence of the pressure increase on the reaction rate is relatively small for simple molecules compared with the influence of temperature. The influence of pressure is dependent on changes in the activation volume. However, the reaction behavior of functional organic molecules is sometimes sensitive to pressure because the biochemical functions of such molecules, such as long peptides and RNA, are based on the construction of three-dimensional structures. In these structures, the number of water molecules associated with the biomolecule changes during the transition state, thus resulting in an activation volume change (Kaddour et al. 2011, 2021; Maurel et al. 2020). This situation is normally observed in the deep ocean, where the pressure reaches approximately 100 MPa at a depth of ~10,000 m. Naturally, higher pressure conditions appear in the crust; therefore, chemical reactions in liquid water in the deep crust are potentially exposed to environments of extremely high pressures.

8.2.3. *Biochemical interactions*

Biochemical studies are widely conducted at low temperatures and pressures because modern organisms survive at relatively low temperatures and pressures. Thus, the literature on the interactions and characteristics of bioorganic molecules at low temperatures and pressures is extensive. Nonetheless, biologically important interactions, such as hydrogen bonding and hydrophobic interactions, should also work in hyperthermophilic organisms at high temperatures and pressures. In modern organisms, biochemical reactions are controlled by stereospecific interactions between enzymes and ribozymes, which are dependent on weak interactions. However, studies on such interactions under high temperatures and pressures are limited.

Liquid water possesses characteristics that are regarded as unique compared with other liquid materials. Thus, focus must be placed on the interactions that appear, particularly in liquid water. In general, electrostatic interactions, hydrophobic interactions (and occasionally $\pi-\pi$ stacking) and hydrogen bonding are considered to be important for the expression of biomolecular structures and functions (Figure 8.7). All these interactions are strongly dependent on hydrogen bonding. In addition, hydrophobic interactions are caused by the structural formation of water molecules via hydrogen bonding. Thus, these interactions are notably affected by changes in both the temperature and dielectric constant of water.

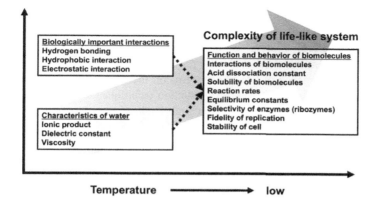

Figure 8.7. *Chemical evolution of life-like system is dependent of the characteristics of water and biologically important interactions*

The ionic product of water increases with increasing temperature up to approximately 250°C and is dependent on pressure (Marshall and Franck 1981). Thus, a neutral pH of 7 with the same amount of protons and hydroxyl ions shifts to lower than 7 at higher temperatures. In addition, the acid dissociation constants of organic molecules, such as amino acids, change with increasing temperature. Thus, the distribution of the charged chemical species formed by protonation and deprotonation changes with increasing temperature. This is effective for the chemical behavior of biologically important molecules because they possess positive and/or negative charges in their structures. Currently, the hydrothermal medium from acidic to alkaline conditions is only sometimes focused on because the direct measurement of pH at high temperatures is still not straightforward. Thus, detailed investigations of how biomolecules react under hydrothermal conditions at different pH values at high temperatures are expected in the future. The decrease in hydrogen bonding and hydrophobic interactions, change in ionic product, and decrease in dielectric constant with temperature, are indicative of the fact that the special characteristics of water are reduced at high temperatures.

Furthermore, the importance of the solubility of biomolecules in relation to chemical evolution has not yet been elucidated. Hence, the solubility of organic molecules under hydrothermal conditions in relation to the origin of life has not yet been widely investigated. The solubilities of electrolytes and biomolecules in water are relatively high or moderate, which is a favorable characteristic of water compared with other solvents. Dissolved materials migrate freely in the aqueous phase, thus allowing smooth interactions among biomolecules.

8.2.4. Minerals and the thermodynamically open system

The hydrothermal system is originated in contact with high-temperature minerals; thus, the minerals should have played important roles in the hydrothermal system for the emergence of life-like systems on the Hadean Earth (Andersson and Holm 2000; Cleaves et al. 2014). The simulation techniques for the roles of minerals must be improved to accommodate actual Hadean environments. In addition, high-pressure reactors are also important as they correspond to environments at higher pressures inside the deep ocean and water in the crust.

The simulation of a thermodynamically open system is another essential aspect. Such a system is observed in real hydrothermal vent systems in the deep ocean (Figure 8.6). Biologically important molecules, such as peptides and nucleic acids, are degraded to shorter monomers in aqueous solutions in a thermodynamically closed system at high temperatures. However, such molecules could have accumulated in thermodynamically open systems as the organisms are maintained by controlling both formation and degradation rates. Thus, the construction of a steady-state or open-flow system is an important issue for the simulation of realistic hydrothermal systems (Figure 8.8). In particular, an open system in the presence of minerals is advantageous and practical for the investigation of realistic hydrothermal systems.

In conclusion, a variety of simulation setups can be developed to investigate hydrothermal conditions. Thus, any type of extremely high temperature and pressure conditions that may have existed on primitive Earth or on other celestial bodies must not be eliminated.

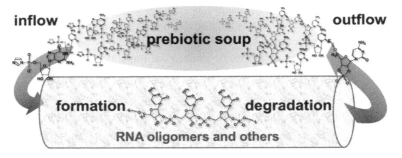

Figure 8.8. Accumulation of oligonucleotides in an RNA-based life-like system is determined by both inflow and outflow rates and formation and degradation of oligonucleotides. For a color version of this figure, see www.iste.co.uk/dimauro/firststeps.zip

8.3. Hydrothermal methodologies regarding the origin-of-life study

8.3.1. Technical background of research tools for hydrothermal reactions

The experimental development of research tools during the last 20–30 years is summarized herein based on the viewpoint of whether the RNA world hypothesis is compatible with the Hadean Earth environments. Both the observation and analysis of natural hydrothermal systems, and the measurement and simulation in laboratories are important strategies for determining whether the chemical evolution reactions proceeded in the ancient Earth. Geological observations, sampling and measurements worldwide can provide some useful evidence for estimating the chemical evolution and ancient organisms (Kitadai et al. 2019). In laboratory simulations and measurements, experimental approaches (Holm and Andersson 2005; Kawamura 2017), where the reaction conditions are simplified and controlled, are useful for investigating the origin of life. Therefore, this reaction has been reproduced by other researchers. However, such conditions may sometimes be too simplified compared with natural hydrothermal systems. Such studies do not reflect any realistic phenomena regarding chemical evolution on primitive Earth. Natural hydrothermal systems are complicated compared with simulation experiments in laboratories. Natural hydrothermal systems allow the design of laboratory experiments, such as dry wet cycles, near hydrothermal systems. In addition, terrestrial freshwater is considered a suitable environment for the formation of biopolymers (Da Silva et al. 2015; Damer and Deamer 2020).

A fluid beyond the critical point is called a supercritical fluid. Although supercritical water may be a potentially important fluid medium where chemical evolution could have occurred, such investigations in relation to the origin of life are limited. To simulate hydrothermal conditions, special techniques are required for running reactions at high temperatures and pressures. Traditionally, high-temperature and pressure-resistant vessels (batch reactors) have been used to investigate reactions. A batch reactor, sometimes called an autoclave, is resistant to high temperatures and pressures up to approximately the critical point of water. Initially when studying hydrothermal reactions in relation to the origin of life, some difficulties were encountered while performing reactions regarding the chemical evolution of biomolecules using batch reactors. For example, the stability of RNA and peptides was not known in detail, and monitoring at the second to millisecond time scale was necessary at high temperatures (White 1984; Kawamura 1998, 1999, 2000). Note that the heating time of the batch reactor is a few minutes because of heating the batch reactor vessel itself. Therefore, rapid reactions cannot be monitored using batch reactors.

8.3.2. Recent development using flow system

Thus, at the end of the 1990s, flow reactors were used in origin-of-life studies to overcome the aforementioned limitations. Imai et al. (1999) and Kawamura (1998) began independently building hydrothermal flow systems for origin-of-life studies of biomolecules in 1998. Originally, such hydrothermal flow systems were used in both fundamental and practical areas, rather than in origin-of-life studies. Normally, the vapor pressure of water increases with increasing temperature; therefore, running at higher temperatures requires higher mechanical resistance. Furthermore, the reactor vessel materials are exposed to highly collusive water at high temperatures, which results in the dissolution and oxidation of the vessel materials. Technical knowledge of resistance to hydrothermal conditions has been accumulated by investigations using conventional vessels, particularly in the fields of inorganic chemistry and metal materials. Imai et al. proposed an experimental setup that simulates a realistic hydrothermal circulation system in the deep ocean (Figure 8.6). The general components of hydrothermal flow reactor systems, including simulation and measurement experiments, can be summarized in relation to the concept of hydrothermal reactor systems for origin-of-life studies (Figure 8.9). Currently, different types of research tools are utilized in studies investigating the origin of life (Mitsuzawa et al. 2005).

First, the flow system enables the simulation of circulation in realistic hydrothermal systems in the deep ocean. The concept of circulation reported by Imai et al. (1999) is based on a nonequilibrium system and is applicable to various realistic simulations of ancient environments on Earth. This type of simulation of hydrothermal reactions using a flow reactor can be regarded as analogous to Miller's experiment. Hydrothermal reactions have been conducted by numerous researchers (Islam et al. 2003; Cleaves et al. 2009). Recently, the roles of alkaline hydrothermal systems were simulated to form short oligonucleotides from the activated nucleotide monomer (Herschy et al. 2014). Our group developed measurement methods using a very narrow tubing reactor with an inner diameter of 0.015–0.25 mm for monitoring reactions within 0.002–200 s at 30 MPa and 400°C. This study focused primarily on obtaining kinetic data by changing the residence time of the samples exposed to high temperatures and pressures (Kawamura, 1998, 1999, 2000, 2017). This method has been continuously improved and applied for different purposes, as summarized in Table 8.1. An absorption spectrophotometric detector (Kawamura 2002; Kawamura et al. 2010) and the attachment of the reaction column packed with mineral particles are shown in Figure 8.10 (Kawamura et al. 2011). These instruments enable the in situ monitoring of hydrothermal reactions in the UV–Vis–NIR spectrophotometric range of 200–2200 nm, in the presence and absence of different minerals (Kawamura et al. 2017, 2019). The difficulty in monitoring the absorption spectra for rapid reactions was overcome using the flow system, which enables the monitoring of rapid hydrothermal reactions at temperatures of up to 400°C within the timescale of 0.8–60 s.

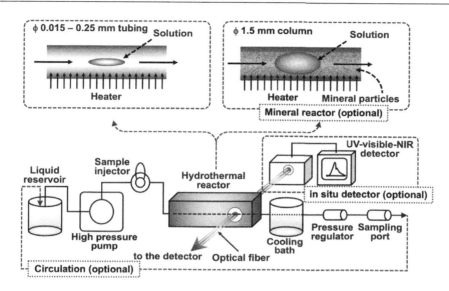

Figure 8.9. *General schematic diagram of hydrothermal flow systems including an absorption spectrophotometric detector and/or a mineral packed reactor column. For a color version of this figure, see www.iste.co.uk/dimauro/firststeps.zip*

Flow reactors	Purpose and improvement	References
Real-time monitoring	Batch reactor to flow reactor for observation of rapid reactions	Kawamura et al.1998, 1999, 2000
Simulation of submarine vent system	Simulated circulation with flow reactor	Imai et al. 1999; Islam et al. 2003; Cleaves et al. 2009
In situ monitoring	Attachment with absorption spectrophotometer for flow system	Kawamura et al. 2003, 2010
Mineral-mediated flow reactor	Attachment with mineral particle packed column for flow system	Kawamura et al. 2011
Mineral-mediated flow reactor with in situ monitoring	Attachment with UV–Vis–NIR absorption spectrophotometer and mineral particle packed column for flow system	Kawamura et al. 2017, 2019
High-pressure flow reactor	High-pressure measurement for ribozymes	Kawamura et al. 2022

Table 8.1. *Different types of flow reactor systems for hydrothermal origin-of-life study*

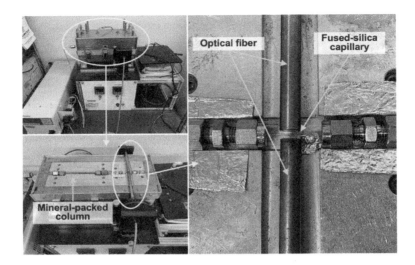

Figure 8.10. *Spectrophotometric detector device with the mineral packed reactor column for the hydrothermal reactor (Kawamura et al. 2017). For a color version of this figure, see www.iste.co.uk/dimauro/firststeps.zip*

With this type of spectrophotometric system, the kinetic and thermodynamic parameters can be obtained under hydrothermal conditions as using conventional spectrophotometers at low temperatures. Conventional absorption spectrophotometers, which are useful for obtaining thermodynamic data at low temperatures, frequently use an optional adaptor for measurements at different temperatures. However, the temperature limit is considerably lower than 100°C. In addition, conventional spectrophotometry with high-temperature attachment does not allow the rapid monitoring of hydrothermal reactions. The series of spectrophotometric detection systems used in our study enabled the measurement of absorption spectra under hydrothermal conditions. In addition, the hydrothermal flow reactor was applied as a high-pressure research tool of up to 30 MPa, which was applied to a ribozyme of avocado viroid (Kawamura et al. 2022b).

8.4. RNA world versus hydrothermalism

8.4.1. *Stability and accumulation of RNA*

Hydrothermal conditions are considered suitable for the formation of simple molecules. Building blocks, such as amino acids, nucleobases, and sugar, can be formed in hydrothermal environments (Hennet and Holm 1992; Holm and Andersson 2005; Martin et al. 2008; Kopetzki and Antonietti 2011; Westall et al. 2018). The stepwise chemical evolution of RNA is summarized in a previous publication (Kawamura and Maurel 2017). On the contrary, hydrothermal environments do not appear to be very suitable for the formation of oligo- or poly-peptides and nucleotides. The stability of biomolecules is an important factor in evaluating whether the functional molecules survived under Hadean Earth conditions, as well as the formation behaviors of biomolecules. Note that the accumulation of biomolecules is determined by both the reaction rates of formation and degradation (Figure 8.8). Investigations on the stability of biomolecules have been traditionally performed using conventional batch reactors and flow reactors (Kawamura 2000, 2003a, 2003b; Kawamura and Yukioka 2001). The kinetics of the stability of these molecules is systematically shown herein (Figure 8.11). Numerous organic molecules in organisms, particularly peptides and nucleic acids, are readily hydrolyzed in aqueous conditions with increasing temperatures. In particular, long-chain nucleic acids and peptides have a higher probability of breaking down at the phosphodiester and amide bonds by hydrolytic degradation.

According to these investigations, peptides and nucleic acids are degraded within a minute to a second at 300°C. Peptide bonding is much more stable than phosphodiester bonding in RNA; however, this does not reflect whether peptides or RNA could have accumulated easily under hydrothermal conditions. The stability of peptides is evaluated by the racemization of the molecules because the racemization process is fairly fast, and deamination and decarboxylation are not detected within a reasonable time period (Kawamura and Yukioka 2001). This may be because racemization proceeds via an ionic process by deprotonation at the alpha carbon of the chiral center, whereas deamination and decarboxylation involve bond-breaking processes. By contrast, the hydrolytic degradation of phosphodiester bonds within RNA and DNA is a determining factor for the stability of RNA and DNA. DNA is more stable than RNA at temperatures up to approximately 200°C; however, this stability is reversed at higher temperatures. This is because the enthalpy change for the phosphodiester bond in DNA is higher than that in RNA, whereas the entropy change for DNA is much higher than that for RNA. Thus, the apparent free energy change decreases with increasing temperature for DNA (Kawamura 2003a). The impression that DNA is stable and presumably suitable as a genetic material is true at low temperatures; however, this perception is not applicable under hydrothermal conditions beyond approximately 200°C. The stability of a hammerhead ribozyme was examined by using hydrothermal flow reactor indicating that the ribozyme is very unstable under hydrothermal conditions (El-Murr et al. 2012).

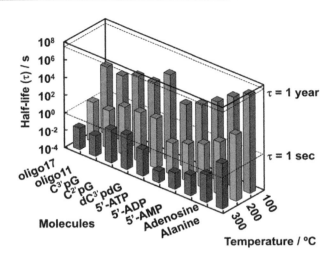

Figure 8.11. *Stability of oligonucleotides, dinucleotides, nucleotides, nucleotide base and amino acid in aqueous solution at high temperatures (Kawamura and Yukioka 2001, Kawamura 2000, 2003a, 2003b). Oligo17: 5'-GGCCGGTTTTCCGGCC, oligo11: 5'-GGCCGGTTTTT (underlines indicate ribose phosphodiester bond). For a color version of this figure, see www.iste.co.uk/dimauro/firststeps.zip*

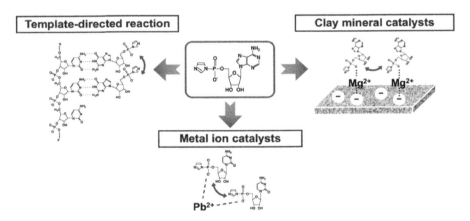

Figure 8.12. *Primitive RNA formation models that were mainly examined under mild conditions (Sawai 1974, Inoue and Orgel 1982, Ferris and Ertem 1992). For a color version of this figure, see www.iste.co.uk/dimauro/firststeps.zip*

The fate of functional RNA molecules changes if the molecules settle under nonequilibrium conditions. This can be seen in organisms where formation and degradation are controlled rapidly by a set of enzymes, even though higher molecules are degraded to lower molecules under the thermodynamically isolated and closed system. Currently, techniques for nonequilibrium simulations under hydrothermal conditions are not well established in laboratories. The possible accumulation of higher RNA molecules under hydrothermal conditions can be estimated based on limited knowledge (Kawamura 2004, 2012, 2019).

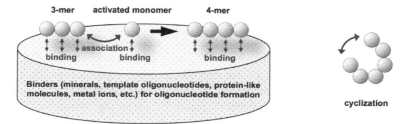

Figure 8.13. *Phosphodiester bond formation between oligonucleotide and monomer on binders (left) and for cyclization (Kawamura and Umehara 2001; Kawamura and Maeda 2007, 2008; Kawamura et al. 2003). For a color version of this figure, see www.iste.co.uk/dimauro/firststeps.zip*

First, accumulation is possible if the formation rate is greater than the degradation rate under hydrothermal conditions. To evaluate this, we analyzed why RNA oligomer formation is not easy at higher temperatures by comparing the formation and degradation rates using prebiotic formation models, as shown in Figure 8.12 (Sawai 1976; Inoue and Orgel 1982; Ferris and Ertem 1992). These investigations suggest that the degradation of formed oligomers is not the reason for the low efficiency of oligonucleotide formation at higher temperatures (Kawamura and Umehara 2001; Kawamura and Maeda 2007, 2008; Kawamura et al. 2003). The real reason assumed was that the rate of phosphodiester bond formation is relatively low compared with that of the degradation of the activated monomers. This is primarily because the association between the activated monomer and elongating oligomer weakens at higher temperatures (Figure 8.13, left). Essentially, the overall reaction rate depends on the association step. This is consistent with cyclization by forming a phosphodiester bond (Figure 8.13, right). The formation of phosphodiester bonds by cyclization is advantageous for the formation of associates prior to the formation of phosphodiester bonds. This is consistent with the chemical reason that the enzyme suppresses the freedom of the reactants by settling molecules with elaborate stereospecificity. These analyses suggest that the formation and degradation rates become competitive at approximately 160–380°C. Second, the long peptides and nucleic acids formed are quenched immediately in cool water (Figure 8.6). For peptides, relatively effective processes for forming oligopeptides from amino acid monomers or short oligopeptides have been reported (Imai et al. 1999; Kawamura et al. 2005; Kawamura and Shimahashi 2008). However, the formation processes of long RNA have not yet been discovered in aqueous solutions under hydrothermal environments.

8.4.2. RNA-based life-like system under hydrothermal environments

In modern organisms, biochemical reactions are controlled by enzymes at a very high speed (Radzicka and Wolfenden 1995; Weber 2004; Kawamura 2010, 2019). This guarantees the selectivity of the enzymes and rate of evolution. This idea is important because the rate of biological evolution depends on the reaction rate. In addition, the accuracy of enzymatic reactions including ribozymes at high temperatures is low compared with that at low temperatures owing to the decrease in biologically weak interactions. These conditions determine the accuracy and fidelity of enzymatic chemical networks, including ribozymes, where temperature determines both the enzymatic reaction rate and interactions between the enzyme and substrate. In such chemical networks, primitive ribozymes maintain both the gap between formation and degradation rate, and biologically weak interactions. However, the cell type compartment is essential for modern organisms. This could have been adapted at an early stage for the emergence of life-like systems (Hogeweg and Takeuchi 2003; Hanczyc et al. 2003; Szostak et al. 2016; Joyce and Szostak 2018).

Weak interactions, such as hydrogen bonding and hydrophobic interactions, are important in determining the fidelity of enzymes and emergence of cell-type compartments. Presumably, this factor primarily determines the temperature limit of primitive life-like systems. Previously, the association of bovine serum albumin (BSA) with chromogenic reagents was determined using a hydrothermal spectrophotometric system (Kawamura et al. 2010) to evaluate the biologically weak interactions at high temperatures. Reportedly, the association of BAS and a chromogenic reagent at 100°C was found to become 1/100 of that at 25°C, but it was still active. In addition, the formation of the double helix of DNA associated with an intercalator was found to become weak at temperatures beyond 100°C. Possibly, the contribution of electrostatic interactions controlling the three-dimensional structure increases with increasing temperature. These facts imply that the weakness of the association between biomolecules at high temperatures results in low fidelity of the life-like system. In addition, the low fidelity can provide a higher possibility of mutation during replication by RP-ribozymes, thus providing different ribozymes.

Although minerals play an important role in the origin-of-life processes, investigations on the formation of RNA molecules in the presence of minerals under hydrothermal conditions are limited. The hypothesis that a mineral surface could have played a similar role in the integration of organic molecules prior to the cell-type compartment, as proposed in our previous study, is currently being evaluated (Kawamura 2012b). Functional RNA molecules adsorb onto the mineral surface and act as integrated chemical networks. Recently, the biochemical behavior of ribozymes on mineral surfaces has been investigated by various research groups (Kawamura et al. 2022a; Takahashi and Sugimoto 2023). Although the measurements were performed at relatively low temperatures compared with the real hydrothermal vent system, the biochemical reaction of the hammerhead ribozyme was active at 20–60°C under 30–100 MPa. The assumption that the interaction of biomolecules on the surface also becomes weak at high temperatures should be evaluated experimentally by novel in situ monitoring of interactions on mineral surfaces under hydrothermal conditions.

Although information regarding the assumption of the upper limit temperature is insufficient, the limit temperature is thought to be higher than the highest temperature at which hyperthermophilic organisms survive. For instance, the optimum temperature of artificial hyperthermophilic enzymes can be extended to higher temperatures by designing artificial proteins (Akanuma et al. 2013). In addition, according to our previous comparison between prebiotic degradation and RNA formation, the limit of possible temperature must be over 300°C, where the formation and degradation rates become competitive (Kawamura and Umehara 2001; Kawamura and Maeda 2007, 2008). Presumably, an RNA-based life-like system under hydrothermal conditions would be possible if the chemical network was associated with an enhancer to support the association between ribozymes and other molecules (Figure 8.13). Under the same principle, which was applied to temperature as a physical factor for life, the factor of pressure would be considered; however, no simple relationship between the reaction rate and pressure, and the pressure change and chemical evolution are expected.

Based on these facts, chemical evolution, including RNA, could have experienced a balanced trajectory between the advantage of the rapid reaction and the disadvantage of the low accuracy of enzymatic reactions (Figures 8.14 and 8.7). In conclusion, the hypothesis that the trajectory between the temperature drop and the proficiency of an RNA-based life-like system is synchronized on the early Earth was briefly discussed previously (Kawamura 2004, 2012b, 2019). A possible trajectory between the chemical evolution and the decrease in temperature on Earth is illustrated in Figure 8.14. The chemical functions of RNA molecules are dependent on temperature and pressure, and weak interactions are dependent on temperature. The temperature on Earth should also have decreased during chemical evolution. The assignment between information and function using RNA molecules was replaced by the assignment method including proteins and DNA. Compartment methods have gradually evolved from a simple integration mechanism to the cell type compartment, where mineral surfaces support the integration of RNA reactions as a candidate integration mechanism. Although the integration of functional RNA molecules on minerals may be easier than in the cell type compartment, this should be evaluated using a hydrothermal reactor system. These important events during chemical evolution could have synchronized with a decrease in temperature. This hypothesis implies that the upper limit temperature for an RNA-based life-like system can be identified by analyzing ribozyme functions in the presence of minerals under hydrothermal conditions. The strategy to identify possible RNA-based life-like molecules is shown in Figure 8.15. Nonetheless, technical improvements in hydrothermal studies are limited.

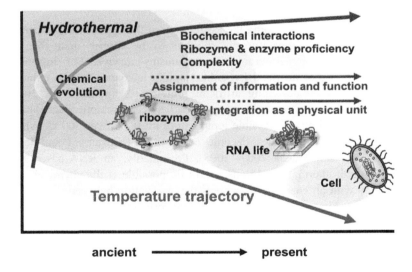

Figure 8.14. *A possible trajectory between the chemical evolution and the decrease in temperature on Earth. For a color version of this figure, see www.iste.co.uk/dimauro/firststeps.zip*

8.5. Future outlook and conclusions

Hydrothermal environments involve a variety of chemical and physical conditions, a wide range of temperatures and pressures, time ranges and other chemical conditions in relation to the presence of different minerals and dissolved materials. Future approaches for the investigation of RNA-based life-like systems from the viewpoint of hydrothermalism are illustrated in Figure 8.15. This chapter has demonstrated that the development of research tools is key to advancing the origin-of-life studies. However, some technical difficulties still exist while investigating the origin of life under hydrothermal conditions. First, although the pH and oxidation conditions are crucial, our systems cannot be easily applied to such electrode techniques. This is because the pH and electrochemical measurements are highly sensitive to the flow system, thus resulting in fluctuations in the electrical double layer. Second, the simulation of hydrothermal vent systems in the deep ocean primarily requires very long-term circulation of water between the ocean and crust. Naturally, circulation simulations can provide valuable information for unknown long-term processes. However, the real processes in hydrothermal vent systems take a few years to decades, and such a long-term circulation is not easily simulated in a laboratory experimental setup.

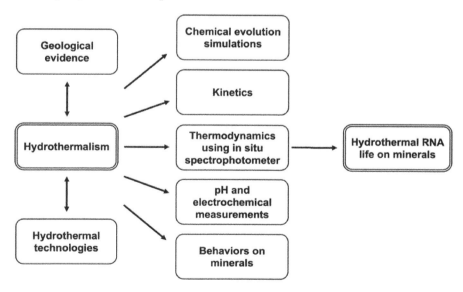

Figure 8.15. *Future approach for hydrothermalism on the origin-of-life study*

Numerous hydrothermal studies have focused on determining the style to know the types of organic processes that could occur under such conditions. Through continuous extensive studies on the formation and observation of biomolecules, our knowledge on the presence and formation process of biomolecules from primitive inorganic materials is increasing. However, how and where relatively simple biomolecules gained biologically important functions, such as information translation and metabolic networks, are unknown. The hydrothermal environment, as an initial chemical environment in which simple biomolecules constructed a united chemical network and established primitive biological informational flow and metabolism, has not been sufficiently studied. Our future goal is to empirically evaluate the possibility of hydrothermal conditions as an RNA-based life-like system, which can provide suitable chemical environments for the formation of the most primitive life-like system based on RNA molecules.

8.6. Acknowledgments

This study was supported by JSPS KAKENHI Grant Numbers JP19H02017 (Grant-in-Aid for Scientific Research on Innovative Areas) in 2019–2023. The author would like to express their gratitude to Professor Ernesto Di Mauro for the invitation to contribute to this book. The author would also like to thank Editage (www.editage.com) for their English language editing.

8.7. References

Akanuma, S., Nakajima, Y., Yokoboria, S., Kimura, M., Nemoto, N., Mase, T., Miyazono, K., Tanokura, M., Yamagishia, A. (2013). Experimental evidence for the thermophilicity of ancestral life. *Proc. Natl. Acad. Sci. USA*, 110, 11067–11072.

Andersson, E. and Holm, N.G. (1992). The stability of some selected amino acids under attempted redox constrained hydrothermal conditions. *Origins Life Evol. Biosphere*, 30, 9–23.

Bada, J.L. and Korenaga, J. (2018). Exposed areas above sea level on Earth >3.5 gyr ago: Implications for prebiotic and primitive biotic chemistry. *Life*, 8, 55.

Bolli, M., Micural, R., Eschenmoser, A. (1997). Pyranosyl-RNA: Chiroselective self-assembly of base sequences by ligative oligomerization of tetranucleotide-2',3'-cyclophosphates (with a commentary concerning the origin of biomolecular homochirality). *Chem. Biol.*, 4(4), 309–320.

Brack, A. (1993). Liquid water and the origin of life. *Origins Life Evol. Biosphere*, 23, 3–10.

Brock, T.D. (1985). Life at high temperatures. *Science*, 230, 132–138.

Cech, T.R., Zaung, A.J., Grabowski, P.J. (1981). In vitro splicing of the ribosomal RNA precousor of Tetrahymena: Involvement of a guanosine nucleotide in the excision of the intervening sequence. *Cell*, 27, 487–496.

Cleaves, H.J., Aubrey, A.D., Bada, J.L. (2009). An evaluation of the critical parameters for abiotic peptide synthesis in submarine hydrothermal systems. *Origin Life Evol. Biosphere*, 39, 109–126.

Cleaves, H.J., Scott, A.M., Hill, F.C., Leszczynski, L., Sahaide, N., Hazen, R. (2014). Mineral–organic interfacial processes: Potential roles in the origins of life. *Chem. Soc. Rev.*, 41, 5502–5525.

Costanzo, G., Saladino, R., Crestini, C., Ciciriello, F., Di Mauro, E. (2007). Nucleoside phosphorylation by phosphate minerals. *J. Biol. Chem.*, 282, 16729–16735.

Crick, F. (1970). Central dogma of molecular biology. *Nature*, 227, 561–563.

Da Silva, L., Maurel, M.-C., Deamer, D. (2015). Salt-promoted synthesis of RNA-like molecules in simulated hydrothermal conditions. *J. Mol. Evol.*, 80, 86–97.

Damer, B. and Deamer, D. (2020). The hot spring hypothesis for an origin of life. *Astrobiology*, 20(4), 429–452.

Eigen, M. (1971). Self-organization of matter and the evolution of biological macromolecules. *Naturwissenschaften*, 58, 465–523.

Eigen, M. and Shuster, O. (1978). The hypercycle. A principle of natural self-organization. Part B: The abstract hypercycle. *Naturwissenschaften*, 65, 7–41.

El-Murr, N., Maurel, M.-C., Rihova, M., Vergne, J., Hervé, G., Kato, M., Kawamura. K. (2012). Behavior of a hammerhead ribozyme in aqueous solution at medium to high temperatures. *Naturwissenschaften*, 99, 731–738.

Ellington, A.D. and Szostak, J.W. (1990). In vitro selection of RNA molecules that bind specific ligands. *Nature*, 346, 818–822.

Ferris, J.P. and Ertem, G. (1992). Oligomerization of ribonucleotides on montmorillonite: Reaction of the 5'-phosphorimidazolide of adenosine. *Science*, 257, 1387–1389.

Fox, S.W. and Harada, K. (1958). Thermal copolymerization of amino acids to a product resembling protein. *Science*, 128, 1214.

Gilbert, W. (1986). Origin of life: The RNA world. *Nature*, 319, 618.

Hanczyc, M.M., Fujikawa, S.M., Szostak, J.W. (2003). Experimental models of primitive cellular compartments: Encapsulation, growth, and division. *Science*, 302, 618–622.

Harrison, T.M. (2009). The Hadean crust: Evidence from >4 Ga Zircons. *Ann. Rev. Earth Planet Sci.*, 37, 479–505.

Hennet, R.J. and Holm, N.G. (1992). Abiotic synthesis of amino acids under hydrothermal conditions and the origin of life: A perpetual phenomenon? *Naturwissenschaften*, 79, 361–365.

Herschy, B., Whicher, A., Camprubi, E., Watson, C., Dartnell, L., Ward, J., Evans, J.R.G., Lane, N. (2014). An origin-of-life reactor to simulate alkaline hydrothermal vents. *J. Mol. Evol.*, 79, 213–227.

Hogeweg, P. and Takeuchi, N. (2003). Multilevel selection in models of prebiotic evolution: Compartments and spatial self-organization. *Origin Life Evol. Biosphere*, 33, 375–403.

Holm, N.G. and Andersson, E. (2005). Hydrothermal simulation experiments as a tool for studies of the origin of life on Earth and other terrestrial planets: A review. *Astrobiology*, 5(4), 444–460.

Horning, D.P. and Joyce, G.F. (2016). Amplification of RNA by an RNA polymerase ribozyme. *Proc. Natl. Acad. Sci. USA*, 113, 9786–9791.

Hsu, H.-W., Postberg, F., Sekine, Y., Shibuya, T., Kempf, S., Horányi, M., Juhász, A., Altobelli, N., Suzuki, K., Masaki, Y. et al. (2015). Ongoing hydrothermal activities within Enceladus. *Nature*, 515, 207–210.

Ikehara, K. (2009). Pseudo-replication of [GADV]-proteins and origin of life. *Int. J. Mol. Sci.*, 10, 1525–1537.

Imai, E., Honda, H., Hatori, K., Brack, A., Matsuno, K. (1999). Elongation of oligopeptides in a simulated submarine hydrothermal system. *Science*, 283, 831–833.

Inoue, T. and Orgel, L.E. (1982). Oligomerization of (guanosine 5'-phosphor) -2-methylimidazolide on poly(C): An RNA polymerase model. *J. Mol. Biol.*, 162, 201–217.

Islam, M.N., Kaneko, T., Kobayashi, K. (2003). Reaction of amino acids in a supercritical water-flow reactor simulating submarine hydrothermal systems. *Bull. Chem. Soc. Japan*, 76, 1171–1178.

Johnston, W.K., Unrau, P.J., Lawrence, M.S., Glasner, M.E., Bartel, D.P. (2001). RNA-catalyzed RNA polymerization: Accurate and general RNA-templated primer extension. *Science*, 292, 1319–1325.

Joyce, G.F. (2002). The antiquity of RNA based evolution. *Nature*, 418, 214–221.

Joyce, G.F. and Szostak, J.W. (2018). Protocells and RNA self-replication. *Cold Spring Harbor Per. Biol.*, 10(9), a034801.

Kaddour, H., Vergne, J., Guy Hervé, G., Maurel, M.-C. (2011). High-pressure analysis of a hammerhead ribozyme from Chrysanthemum chlorotic mottle viroid reveals two different populations of self-cleaving molecule. *FEBS J.*, 278, 3739–3747.

Kaddour, H., Lucchi, H., Hervé, G., Vergne, J., Maurel, M.-C. (2021). Kinetic study of the avocado sunblotch viroid self-cleavage reaction reveals compensatory effects between high-pressure and high-temperature: Implications for origins of life on Earth. *Biology*, 10, 720.

Kashefi, K. and Lovley, D.R. (2003). Extending the upper temperature limit for life. *Science*, 301, 934.

Kasting, J.F. (1993). Earth's early atmosphere. *Science*, 259, 920–926.

Kasting, J.F. (1997). Habitable zones around low mass stars and the search for extraterrestrial life. *Origins Life Evol. Biosphere*, 27, 291–307.

Kawamura, K. (1998). Kinetic analysis of hydrothermal reactions by flow tube reactor – Hydrolysis of adenosine 5'-triphosphate at 398–573 K. *Nippon Kagaku Kaishi*, 1998, 255–262.

Kawamura, K. (1999). Monitoring of hydrothermal reactions in 3 ms using fused-silica capillary tubing. *Chem. Lett.*, 28, 125–126.

Kawamura, K. (2000). Monitoring hydrothermal reactions on the millisecond time scale using a micro-tube flow reactor and kinetics of ATP hydrolysis for the RNA world hypothesis. *Bull. Chem. Soc. Japan.*, 73, 1805–1811.

Kawamura, K. (2002). In situ UV-VIS detection of hydrothermal reactions using fused-silica capillary tubing within 0.08–3.2 s at high temperatures. *Anal. Sci.*, 18, 715–716.

Kawamura, K. (2003a). Kinetics and activation parameter analyses of hydrolysis and interconversion of 2',5'- and 3',5'-linked dinucleoside monophosphate at extremely high temperatures. *Biochim. Biophys. Acta*, 1620, 199–210.

Kawamura, K. (2003b). Kinetic analysis of cleavage of ribose phosphodiester bond within guanine and cytosine rich oligonucleotides and dinucleotides at 65–200°C and its implications on the chemical evolution of RNA. *Bull. Chem. Soc. Japan*, 76, 153–162.

Kawamura, K. (2004). Behavior of RNA under hydrothermal conditions and the origins of life. *Inter. J. Astrobiol.*, 3, 301–309.

Kawamura, K. (2010). Temperature limit for the emergence of life-like system deduced from the pre-biotic chemical kinetics under the hydrothermal conditions. In *Proceedings of the Twelfth International Conference on the Simulation and Synthesis of Living Systems*, Odense.

Kawamura, K. (2012a). Drawbacks of the ancient RNA-based life-like system under primitive Earth conditions. *Biochimie*, 94(7), 1441–1450.

Kawamura, K. (2012b). Reality of the emergence of life-like systems from simple prebiotic polymers on primitive Earth. In *Genesis – In the Beginning: Precursors of Life, Chemical Models and Early Biological Evolution*, Seckbach, J. and Gordon, R. (eds). Springer, Dordrecht.

Kawamura, K. (2016). A hypothesis: Life initiated from two genes, as deduced from the RNA world hypothesis and the characteristics of life-like systems. *Life*, 6(3), 29.

Kawamura, K. (2017). Hydrothermal microflow technology as a research tool for origin-of-life studies in extreme Earth environments. *Life*, 7(4), 37.

Kawamura, K. (2019). A non-paradoxical pathway for the chemical evolution toward the most primitive RNA-based life-like system. In *Evolution, Origin of Life, Concepts and Methods*, Pontarotti, P. (ed.). Springer, Cham.

Kawamura, K. and Maeda, J. (2007). Kinetic analysis of oligo(C) formation from the 5'-monophosphorimidazolide of cytidine with Pb(II) ion catalyst at 10–75°C. *Origins Life Evol. Biospheres*, 37(2), 153–165.

Kawamura, K. and Maeda, J. (2008). Kinetics and activation parameter analysis for the prebiotic oligo-cytidylate formation on Na^+-montmorillonite at 0–100°C. *J. Phys. Chem. A*, 112, 8015–8023.

Kawamura, K. and Maurel, M.-C. (2017). Walking over 4 Gya: Chemical evolution from photochemistry to mineral and organic chemistries leading to an RNA world. *Orig. Life Evol. Biopsh.*, 47, 281–296.

Kawamura, K. and Shimahashi, M. (2008). One-step formation of oligopeptide-like molecules from Glu and Asp in hydrothermal environments. *Naturwissenschaften*, 95(5), 449–454.

Kawamura, K. and Umehara, M. (2001). Kinetic analysis of the temperature dependence of the template-directed formation of oligoguanylate from the 5'-phosphorimidazolide of guanosine on a poly(C) template with Zn^{2+}. *Bull. Chem. Soc. Japan*, 74(5), 927–935.

Kawamura, K. and Yukioka, M. (2001). Kinetics of the racemization of amino acids at 225–275°C using a real-time monitoring method of hydrothermal reactions. *Thermochim. Acta*, 375, 9–16.

Kawamura, K., Nakahara, N., Okamoto, F., Okuda, N. (2003). Temperature dependence of the cyclization of guanine and cytosine mix hexanucleotides with water-soluble carbodiimide at 0–75°C. *Viva Origino*, 31(4), 221–232.

Kawamura, K., Nishi, T., Sakiyama, T. (2005). Consecutive elongation of alanine oligopeptides at the second time range under hydrothermal condition using a micro flow reactor system. *J. Am. Chem. Soc.*, 127(2), 522–523.

Kawamura, K., Nagayoshi, H., Yao, T. (2010). In situ analysis of proteins at high temperatures mediated by capillary-flow hydrothermal UV-Vis spectrophotometer with a water-soluble chromogenic reagent. *Anal. Chim. Acta*, 667, 88–95.

Kawamura, K., Yasuda, T., Hatanaka, T., Hamahiga, K., Matsuda, N., Ueshima, M., Nakai, K. (2017). In situ UV-VIS spectrophotometry within the second time scale as a research tool for solid-state catalyst and liquid-phase reactions at high temperatures: Its application to the formation of HMF from glucose and cellulose. *Chem. Eng. J.*, 307, 1066–1075.

Kawamura, K., Hatanaka, T., Hamahiga, K., Ueshima, M., Nakai, K. (2019). In situ UV–VIS–NIR spectrophotometric detection system as a research tool for environment-friendly chemical processes. *Environmental Technology & Innovation*, 15, 100410.

Kawamura, K., Lambert, J.-F., Ter-Ovanessian, L.M.P., Vergne, J., Hervé, G., Maurel, M.-C. (2022a). Life on minerals: Binding behaviours of oligonucleotides on zirconium silicate and its inhibitory activity for the self-cleavage of hammerhead ribozyme. *Life*, 12, 1689.

Kawamura, K., Ogawa, M., Konagaya, N., Maruoka, Y., Lambert, J.-F., Ter-Ovanessian, L.M.P., Vergne, J., Hervé, G., Maurel, M.-C. (2022b). A high-pressure, high-temperature flow reactor simulating the Hadean Earth environment, with application to the pressure dependence of the cleavage of avocado viroid hammerhead ribozyme. *Life*, 12, 1224.

Kitadai, N., Nakamura, R., Yamamoto, M., Takai, K., Yoshida, N., Oono, Y. (2019). Metals likely promoted protometabolism in early ocean alkaline hydrothermal systems. *Sci. Adv.*, 5, eaav7848.

Kopetzki, D. and Antonietti, M. (2011). Hydrothermal formose reaction. *New J. Chem.*, 35, 1787–1794.

Korenaga, J. (2021). Was there land on the early Earth? *Life*, 11, 1142.

Lee, D.H., Granja, J.R., Martinez, J.A., Severin, K., Ghadiri, M.R. (1996). A self-replicating peptide. *Nature*, 382, 525–528.

Marshall, W.L. and Franck, E.U. (1981). Ion product of water substance, 0–1000°C, 1–10,000 bars New International Formulation and its background. *J. Phys. Chem. Ref. Data*, 10, 295–304.

Martell, A.E. and Smith, R.M. (1974). *Critical Stability Constants, First Supplement*. Springer, New York.

Martin, W., Baross, J., Kelley, D., Russell, M.J. (2008). Hydrothermal vents and the origin of life. *Nature Rev. Microbiol.*, 6, 805–814.

Maruyama, S., Ikoma, M., Genda, H., Hirose, K., Yokoyama, T., Santosh, M. (2013). The naked planet Earth: Most essential pre-requisite for the origin and evolution of life. *Geosci. Front.*, 4, 141–165.

Maurel, M.-C., Leclerc, F., Hervé, G. (2020). Ribozyme chemistry: To be or not to be under high pressure. *Chem. Rev.*, 120, 4898–4918.

Miller, S.L. (1953). A production of amino acids under possible primitive Earth conditions. *Science*, 117, 528–529.

Mitsuzawa, S., Deguchi, S., Takai, K., Tsujii, K., Horikoshi, K. (2005). Flow-type apparatus for studying thermotolerance of hyperthermophiles under conditions simulating hydrothermal vent circulation. *Deep-Sea Res. I*, 52, 1085–1092.

Mojzsis, S.J., Harrison, T.M., Pidgeon, R.T. (2001). Oxygen-isotope evidence from ancient zircons for liquid water at the Earth's surface 4,300 Myr ago. *Nature*, 409, 178–181.

Nemoto, N. and Husimi, Y. (1995). A model of the virus-type strategy in the early stage of encoded molecular evolution. *J. Theor. Biol.*, 176, 67–77.

Nemoto, N., Miyamoto-Sato, E., Husimi, Y., Yanagawa, H. (1997). In vitro virus: Bonding of mRNA bearing puromycine at the 30-terminal end to the C-terminal end of its encoded protein on the ribosome in vitro. *FEBS Lett.*, 414, 405–408.

Orgel, L.E. (2004). Prebiotic chemistry and the origin of the RNA world. *Crit. Rev. Biochem. Mol. Biol.*, 39, 99–123.

Radzicka, A. and Wolfenden, R. (1995). A proficient enzyme. *Science*, 267, 90–93.

Saladino, R., Botta, G., Pino, S., Costanzoc, G., Di Mauro, E. (2012). Genetics first or metabolism first? The formamide clue. *Chem. Soc. Rev.*, 41, 5526–5565.

Santosh, M., Arai, T., Maruyama, S. (2017). Hadean Earth and primordial continents: The cradle of prebiotic life. *Geosci. Front.*, 8, 309–317.

Sawai, H. (1976). Catalysis of internucleotide bond formation by divalent metal ions. *J. Am. Chem. Soc.*, 98, 7037–7039.

Sleep, N.H. (2010). The Hadean-Archaean environment. *Cold Spring Harb. Perspect. Biol.*, 2, a002527.

Szostak, N., Wasik, S., Blazewicz, J. (2016). Hypercycle. *Plos. Comp. Biol.*, 12(4), e1004853.

Takahashia, S. and Sugimoto, N. (2023). Pressure-temperature control of activity of RNA polymerase ribozyme. *Biophys. Chem.*, 292, 106914.

Takai, K., Nakamura, K., Toki, T., Tsunogai, U., Miyazaki, M., Miyazaki, J., Hirayama, H., Nakagawa, S., Nunoura, T., Horikoshi, K. (2008). Cell proliferation at 122°C and isotopically heavy CH_4 production by a hyperthermophilic methanogen under high-pressure cultivation. *Proc. Natl. Acad. Sci. USA*, 105, 10949–10954.

Tjhunga, K.F., Shokhireva, M.N., Horninga, D.P., Joyce, G.F. (2020). An RNA polymerase ribozyme that synthesizes its own ancestor. *Proc. Natl. Acad. Sci. USA*, 117(6), 2906–2913.

Tuerk, C. and Gold, L. (1990). Systematic evolution of ligands by exponential enrichment: RNA ligands to bacteriophage T4 DNA polymerase. *Science*, 249, 505–510.

Uematsu, M. and Frank, E.U. (1980). Static dielectric constant of water and steam. *J. Phys. Chem. Ref. Data*, 9, 1291–1306.

Weber, A.L. (2004). Kinetics of organic transformations under mild aqueous conditions: Implications for the origin of life and its metabolism. *Origins Life Evol. Biospheres*, 34, 473–495.

Westall, F., Hickman-Lewis, K., Hinman, N., Gautret, P., Campbell, K.A., Bréhéret, J.G., Foucher, F., Hubert, A., Sorieul, S., Dass, A.V. et al. (2018). A hydrothermal-sedimentary context for the origin of life. *Astrobiol.*, 18(3), 259–293.

White, R.H. (1984). Hydrolytic stability of biomolecules at high temperatures and its implication for life at 250°C. *Nature*, 310, 430–432.

Zaher, H.S. and Unrau, P.J. (2007). Selection of an improved RNA polymerase ribozyme with superior extension and fidelity. *RNA*, 13, 1017–1026.

9
Studies in Mineral-Assisted Protometabolisms

Jean-François LAMBERT[1], Louis TER-OVANESSIAN[1,2]
and Marie-Christine MAUREL[2]

[1] *Laboratoire de Réactivité de Surface, LRS, UMR 7197 CNRS,
Sorbonne Université, Paris, France*
[2] *Institut de Systématique, Évolution, Biodiversité (ISYEB), École Pratique des Hautes
Études, Muséum National d'Histoire Naturelle, Sorbonne Université,
Université des Antilles, Paris, France*

9.1. Metabolism, protometabolism and minerals

The window for single or, more probably, multiple life emergences on the Earth is framed by two milestones. The first one is the stabilization of physico-chemical conditions compatible with life (4.2 billion years for a stable hydrosphere). The second milestone is the unambiguous identification of non-dubious fossils (2.7 billion years). If these broad limits are mainly agreed by researchers, much else about the origins of life remains obscure. Some form of selection, acting on prebiotic molecules, must have been involved, as well as open systems, exchanging with the prebiotic environment.

When the first steps and reactivity of life unfolded (Maurel and Décout 1999), a diverse mineral world had already been in existence on Earth for several hundred million years (Hazen 2013). This world provided the setting for the increase in biochemical complexity, and more precisely the region of contact between the minerals and the aqueous solutions containing the first biomolecules (the primordial oceans) – namely, the mineral surfaces.

However, as highlighted by Orgel, "little is known with certainty about the physical environment in which life evolved or about the detailed steps that led from unconstrained abiotic chemistry to the organized complexity of biochemistry" (Orgel 2006). This would plead for prebiotic chemical reactions to be explored systematically in conditions corresponding with plausible Archean Earth environments (as far as we know).

The diversity of environments and possible chemistries led to a wide diversity of origin of life scenarios, giving pride of place to conjectures. However, some experimental work started to confirm or question these hypotheses.

Bernal (1949) was the first to advocate for a definite role of mineral surfaces in the origins of life, especially in concentrating the molecular precursors of biomolecules by adsorption, making their reaction less unlikely.

Later on, experimental studies started to demonstrate that reactions of biomolecules could be carried out on mineral platforms, such as the polymerization of amino acids to oligopeptides on clay minerals (Lahav and Chang 1976), or later the polymerization of activated (Ferris 1996) and unactivated (Hassenkam et al. 2020) nucleotides monophosphates.

In the late 1980s, the subject of protometabolism on mineral surfaces was systematized and brought to attention by the work of Wächtershäuser (1988a), pointedly presented as "a theory of surface metabolism". Concentrating on a particular class of mineral surfaces – metal sulfides such as pyrite, FeS_2 – the author contended that "all the pathways of central and universal metabolism can be retrodictively transformed into archaic pathways" (Wächtershäuser 1992). This is a strong assertion according to which the main reactions of current biochemistry recapitulate, or essentially reproduce, similar reactions that already took place in a geochemical setting before the advent of organisms.

Although partly validated by experimental data (Huber et al. 2003), Wächtershäuser's work remained to a large extent "chemistry on paper" and was strongly criticized. Recapitulating data available at the time, Lazcano and Miller (1999) advocated that "(successful) prebiotic synthetic pathways are all different from the biosynthetic pathways". More recent experimental works, such as the extensive investigations of Patel et al. (2015), lent additional strength to this opinion.

The original protometabolic pathways may have been very different from the current ones, including chemical reactions and precursors that are currently not part of metabolisms as we know them today. Obviously, they could not rely on fixtures of current biology such as enzymes for catalysts, or free energy-rich cofactors. To support these reactions, either small polypeptides were able to induce catalytic properties (Gorlero et al. 2009) (proto-enzymes) or, more likely, mineral surfaces were playing a more important role than today.

This implies that, at some point, a takeover from mineral-based to biomolecules-based chemistry took place: mineral catalysts were supplanted by more efficient enzymes, and the delivery of free energy was systematized by the appearance of cofactors. This takeover may have been concomitant with evolution, selection and incorporation processes among a large diversity of protometabolisms. As suggested by Lazcano and Miller (1999), metabolism as we see it today is a kind of "patchwork assembly of biosynthetic routes" with different origins.

In the last few years, researchers have attempted to take up the challenge set by Orgel: "The demonstration of the existence of a complex, nonenzymatic metabolic cycle, such as the reverse citric acid, would be a major step in research on the origin of life, while demonstration of an evolving family of such cycles would transform the subject" (Orgel 2008). They tried to replicate key parts of biological metabolism in inorganic settings (Ralser 2018; Muchowska et al. 2020). Thus, compelling progress has been made toward duplicating most key steps of the reverse tricarboxylic acids cycle in purely inorganic environments (Preiner et al. 2019). Closing the cycle would represent a first example of duplicating an important component of metabolic complexity, and go a long way to rationalize the capture of CO_2 to make carbon-based skeletons. However, many objections raised by Orgel (side reactions, etc.) remain unsolved (Orgel 2000).

At any rate, biochemistry occurring nowadays should certainly not be considered as a target to be identically reproduced under prebiotic conditions (as it would be denying the diversity of primitive unknown metabolisms and evolutionary steps of selection), but rather viewed as providing potential guidelines able to direct experimental work on minerals. The conditions for primordial syntheses should definitely be compatible with what is known of the geochemical context of the Archean era. Many different pathways have to be tested to understand the diversity of protometabolisms and how they influenced each other in the geochemical context.

We shall now consider in some detail, from the point of view of surface science, how mineral surfaces can influence prebiotic reaction pathways.

9.2. Adsorption on mineral surfaces

The establishment of an interaction between a molecule from an aqueous solution, or from the gas phase, and a solid surface is known as adsorption. Adsorption phenomena have long been studied by physical chemists, electrochemists, geochemists and soil scientists. Since they constitute the first step in any mineral surface-mediated scenario, it is relevant to briefly describe them.

9.2.1. *Adsorption mechanisms*

Adsorption phenomena belong to the realm of chemistry, and the interactions they involve are well known in other fields of chemistry. "Physisorption" simply denotes adsorption through weak intermolecular interactions such as van der Waals interactions. Higher energy adsorption phenomena include the following.

9.2.1.1. *Electrostatic adsorption*

Some important mineral phases consist of lattices bearing regularly arranged electric charges, compensated by highly mobile ions of the opposite charge located in their porosity. They include the layered clay minerals, frequent components of prebiotic scenarios (Kloprogge and Hartman 2022) and microporous zeolites (Smith 1998): both have a negative lattice charge and contain easily exchangeable cations. Conversely, hydrotalcites (Greenwell and Coveney 2006) are layered compounds with a positive lattice charge, containing exchangeable anions.

Other surfaces bear specific moieties, like the silanols (SiOH) on the surface of silica, that can behave as Brönsted acids, losing a proton to the solution, and/or as Brönsted bases, being conversely protonated. Thus, they will develop a positive or negative surface charge, depending in this case on the solution pH.

In both cases, the mineral possesses a surface charge, and consequently an electric potential that reaches a few hundred mV with respect to the bulk solution. Ions of the opposite sign will be retained in a *diffuse layer* – for example, in the vicinity of a negative surface, cations will be concentrated and anions will be excluded. There is then a net adsorption of the cations from an electrolyte solution.

9.2.1.2. *Outer-sphere complex formation*

In addition to long-distance electrostatic adsorption, molecules can also interact with the surface through localized bonds, forming "complexes" (defined in a broad sense) with surface groups. If this involves weaker interactions than covalent bonds, such as hydrogen bonds or dipole-dipole interactions, they are called "outer-sphere complexes". In the case of adsorption from aqueous solutions, colloid chemistry incorporates outer-sphere complexes as a layer of specifically adsorbed molecules or ions situated a few Angstroms away from the surface (*Stern layer*) and has developed methods for macroscopic adsorption data analysis that allow us to quantify this specific adsorption and evaluate its energetics. It is very likely that biomolecules can also be adsorbed as outer-sphere complexes on dry surfaces, that is, at the solid/atmosphere interface: such complexes have been especially well documented on silica (Rimola et al. 2013), but their occurrence is probably much broader. Their characterization at the molecular scale is difficult and time-consuming (Abadian et al. 2022), but may be rewarding (see section 9.2.2). The cooperative effect of several H bonds may result in rather strong adsorption, but it will usually be easily reversible, provided that other molecules capable of H-bonding (which include H_2O!) are present in the environment.

9.2.1.3. Inner-sphere complex formation

Many biomolecules can definitely establish covalent bonds with surface groups, yielding "inner-sphere complexes". A case in point is the condensation of a carboxylic acid group with a surface silanol, to give a "surface ester" (R-CO-O-Si) (El Samrout et al. 2022). This reaction is analogous to the formation of an ester with an alcohol group in organic chemistry. Inner-sphere adsorption is not easily reversible, and organic molecules bound in this way are sometimes said to be "anchored" or "grafted" to the surface.

9.2.2. Adsorption selectivities

Besides intervening as an elementary step in heterogeneous catalysis (see section 9.4), adsorption on surfaces may select some molecules from a complex mixture and thus help understand how order emerged from the primordial soup. A well-known example is the explanation of chirality selection by selective adsorption of one enantiomer on intrinsically chiral surfaces such as those of quartz or calcite (Hazen et al. 2001; see also Chapter 2 of this book). But adsorption selectivity is a much more common phenomenon.

Electrostatic adsorption has often been invoked in an origins of life context, for example to explain the selection of negatively charged molecules, such as the metabolites of the tricarboxylic acid cycle or phosphate-activated biomolecules (Wächtershäuser 1988a; Arrhenius et al. 1997), by positively charged surfaces. Experimentally, electrostatic adsorption selectivity could be observed from mixtures of amino acids such as arginine + glutamic acid: in pH conditions where Arg is positively and Glu negatively charged, the clay mineral montmorillonite strongly selected Arg (Jaber et al. 2014), while hydrotalcite selected Glu.

In outer-sphere adsorption, possible selectivity through the formation of a specific lattice of H bonds with a substrate molecule would be particularly interesting because it would be reminiscent of the way in which enzymes recognize their substrate (not to mention DNA base pairing, etc.). While there does not seem to be any conclusive instances of such a selectivity, some studies seem to suggest that adsorption through specific H-bonding patterns could be a necessary first step in silica-catalyzed prebiotic reactions (Fabbiani et al. 2018). At any rate, molecules temporarily retained by outer-sphere adsorption may easily desorb in later steps and be available for further reaction.

Inner-sphere adsorption appears to be a less likely way of selecting molecules from a complex mixture in solution. Because it is strong and hardly reversible, it would rather tend to remove the molecules that undergo this fate from the "playing field" of subsequent transformations. While this may be interesting in specific scenarios, it seems to be less generally applicable than the previous possibility.

Adsorption selectivity can discriminate between different substrates with high efficiency, to the point that molecular recognition has often been invoked, especially in the field of biomedical devices (Castner and Ratner 2002). Even though these generally involve highly engineered surfaces that are not realistic in a geochemical context (Slowing et al. 2007), molecular recognition has been reported even on raw silica surfaces (Boujday et al. 2003).

9.3. Mineral surfaces and reaction thermodynamics

Since the formation of high-energy biomolecules has been observed in the presence of minerals, the latter must modify the thermodynamics of chemical reactions in some way. This may be by taking part directly in the chemical reaction, or by modifying the reagents through adsorption.

9.3.1. *Minerals as reagents*

There are some instances in which minerals act as stoichiometric reagents, being definitively altered by their reactions with organic molecules; if, originally, they were not at equilibrium with their surroundings, part of the free enthalpy of the system may well be captured to form high-energy biomolecules. High-energy mineral agents that could have been introduced into the primordial ocean as a consequence of volcanic activity include COS (Leman et al. 2004) and polyphosphates (Osterberg and Orgel 1972), and particularly the trimetaphosphate ion (Yamagata et al. 1991) often used to jump-start prebiotic transformations (Gibard et al. 2018). Solid-state monophosphate minerals have been observed to successfully phosphorylate nucleosides (Costanzo et al. 2007), but it may be that in these studies the thermodynamic driving force was connected to the use of a non-aqueous solvent (formamide).

Minerals as prebiotic reagents have often been invoked to explain reduction-oxidation reactions. Serpentinization is an important geochemical reaction in which Fe(II) silicates decompose water to produce H_2 in solution. In the very reducing environments formed by serpentinization, it becomes favorable to reduce CO_2 to methane and/or formate (Martin et al. 2007; Martin 2019; Ruiz et al. 2021), in reactions similar to the most ancient metabolic pathways (methanogenesis and acetogenesis). Wächtershäuser proposed that the formation of pyrite ($Fe^{2+}(S_2^{2-})$) from FeS could act as a source of reducing power for early protometabolic reactions (Wächtershäuser 1988b); here, the redox couple involved is $S^{2-}/Fe(S_2)_{solid}$, being drawn to the right by the precipitation of pyrite. Interestingly, it has been shown that amide bonds may form between amino acids and mercaptoacetic acid, even in dilute solutions (see section 9.3.2), in the presence of FeS (Keller et al. 1994); the mechanism was proposed to involve the formation of an activated intermediate (thioacetic acid) that later reacted with the amine moieties. Here, the transfer of free enthalpy from the inorganic to the bioorganic components of the prebiotic system would be demonstrated.

Conversely, minerals could act as oxidizing reagents, a role that was more difficult to fulfill than today in the absence of molecular dioxygen. The Fe^{3+}/Fe^{2+} couple may have played a role; even though Fe^{3+} was much rarer than today, it could be found in minerals such as garnets, pyroxenes, amphiboles, micas or clay minerals (Hazen 2003), and may have been generated from hydrothermal reactions in submarine springs (Russell and Hall 2002).

Many instances have been reported where high-energy organic molecules are formed in the presence of minerals, while the latter do not act as stoichiometric reagents. Several possible explanations have been proposed and are illustrated here for the well-studied case of amino acids polymerization to peptides.

9.3.2. *Concentrating reagents from the solution*

A first effect of adsorption may be to concentrate the reagents in the vicinity of the surface (as in electrostatic adsorption). This would favor all association reactions, that is, those that result in a decrease of the number of species (molecules and ions) in solution: many reactions of anabolism have this property, since a chief role of anabolism is to increase molecular complexity, and thus build molecules with complex skeletons from simpler building blocks.

Generally speaking, concentrating the reaction medium will displace such reactions toward the formation of condensed products. This may easily be understood in terms of Le Châtelier's principle: increasing the concentration displaces the equilibrium in the direction that opposes this constraint, that is, towards condensation, since this decreases the number of molecules in solution. The concentration argument was the original reason Bernal gave for the participation of adsorption in prebiotic reactions.

While this may be true in principle, taking a numerical example will illustrate the limits of this reasoning. Let us consider the polymerization of amino acids to peptides. The simplest instance would be the dimerization of glycine to diglycine:

$$Gly_{aq} + Gly_{aq} \longrightarrow Gly - Gly_{aq} + H_2O$$

This reaction is thermodynamically unfavorable since the associated free enthalpy change is +17.8 kJ/mol at room temperature (Shock 1992), corresponding to an equilibrium constant $K = 2.5.10^{-3}$. The reaction quotient at equilibrium must be equal to K, and with the usual approximations for dilute solutions, this translates to $\frac{[Gly-Gly]}{[Gly]^2} = 2.5.10^{-3}$. This equation can easily be solved to give the percentage of dimerized amino acid as a function of the total concentration (Figure 9.1).

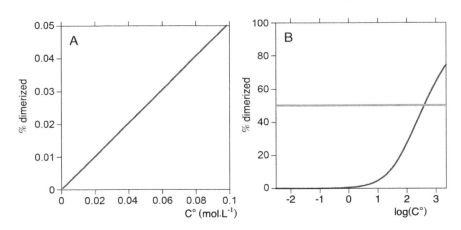

Figure 9.1. Percentage of glycine dimerized at equilibrium as a function of total concentration. (A) is the linear scale and (B) is the logarithmic scale of concentration

It can be seen that the efficiency of dimerization does indeed increase with concentration, almost linearly at low values, but in order to reach 50% dimerization, impossibly high glycine concentrations would have to be attained (~390 mol·L^{-1}). Thus, concentration by adsorption cannot realistically overcome even the relatively small barrier of +17.8 kJ/mol.

9.3.3. *Altering free enthalpies of reaction*

Adsorbed molecules generally do not have the same free enthalpy as molecules in solution – this is already implicit in the existence of concentration effect. If the products of a reaction are more stabilized by adsorption than its reagents, this reaction will have a different free enthalpy change in the *adsorbed phase* compared to the free solution, possibly with a higher value of the equilibrium constant. It is certainly plausible that free enthalpies of adsorption of the reaction products could be negative enough to overcome the unfavorability of many uphill metabolic reactions. For the example of glycine dimerization, this would require adsorption enthalpies in the order of −20 kJ/mol, corresponding to a single strong H bond (outer-sphere complex), while coordinative bonds (inner-sphere complex) may easily provide −50 kJ/mol.

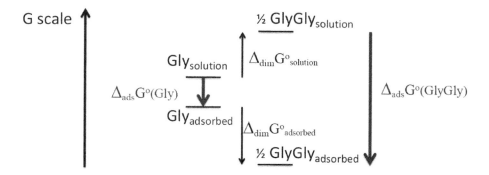

Figure 9.2. *Illustrating de Duve's paradox for the case of glycine polymerization: if peptide growth is favored in the adsorbed phase and unfavored in the solution, each additional monomer makes the peptide chain more strongly adsorbed. For a color version of this figure, see www.iste.co.uk/dimauro/firststeps.zip*

It was quickly realized however (de Duve and Miller 1991) that such a solution to the problem risks inducing an "adsorption trap". If the protometabolic products are much stabilized by adsorption, they will simply remain stuck to the surface. In the words of de Duve and Miller, the (proto)organisms would "evolve into a dead end, in which the surface is covered by molecules so tightly bonded that they cannot be further displaced" (see Figure 9.2).

The best way of course to determine if product stabilization by adsorption really happens is to test it experimentally. It is quite revealing that in the case of amino acids polymerization, this took more than 30 years after the phenomenon was reported. Marshall-Bowman et al. (2010) tried to determine whether glycine oligomerization is more favorable in the presence of minerals such as silica than in the free solution. The answer was negative. As long as an aqueous phase was present, the polymerization equilibria were not measurably affected by adding adsorbing solids to the solution. The efficiency of peptide formation from amino acids must be rationalized in a different way.

9.3.4. *Platforms to capture free energy from macroscopic sources (space gradients and time fluctuations)*

The very first successful attempts at amino acids polymerization (Lahav et al. 1976) applied *wetting-and-drying* cycles to the AA/mineral systems. These cycles were already a standard tool of geochemists to mimic the effect of exposure to changing weather conditions. It is now generally recognized that applying at least one drying treatment at moderately high temperatures (80–160°C) is indispensable to drive the polymerization, not only of amino acids, but also of nucleotides (Higgs 2016). In fact, the rationalization of these findings is quite simple. Still using the case of Gly dimerization as an example, writing the equilibrium constant as $\frac{[Gly-Gly]}{[Gly]^2}$ (section 9.3.2) is based on the hypothesis of working in dilute aqueous solutions. If water activity is not constant (which is certainly the case when a drying step is applied), K should instead be equal to $\frac{a_{Gly-Gly} \cdot a_{H_2O}}{(a_{Gly})^2}$, and if a_{H_2O} decreases upon drying, this must be compensated by an increase in the equilibrium Gly-Gly/Gly ratio to keep K constant. More generally, condensed molecules (such as peptides) will be favored upon drying, as illustrated in Figure 9.3, and since the drying step takes place at relatively high temperatures, we can expect that this new equilibrium state will be reached quickly. If the system is subjected to subsequent rehydration, the condensed molecules will not be at equilibrium anymore, but they will stay frozen in their metastable high-energy state for a long time since the temperature is lower (Higgs 2016; Ross and Deamer 2016) (see Figure 9.3).

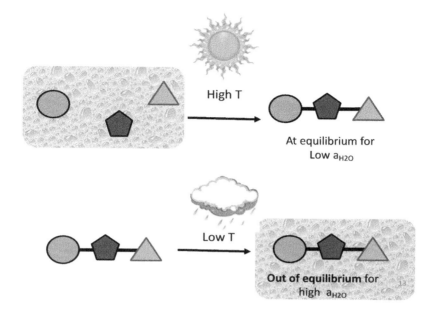

Figure 9.3. *Illustrating the alternation of dry and wet conditions in WD cycles. For a color version of this figure, see www.iste.co.uk/dimauro/firststeps.zip*

In this scenario, the free enthalpy source for the formation of activated (condensed) molecules in solution is ultimately constituted by the macroscopic temporal fluctuations in the environment. This includes humidity cycling, but other cycles, such as of salinity and pH, may have been coupled to it and induced interesting prebiotic phenomena (Ianeselli et al. 2022). In the case of peptides from amino acids and nucleic acids from nucleotides, multiple wetting-and-drying cycles would increase the length of the polymers by allowing their diffusion in the aqueous phase, which would be easier than on the surface. The role of the mineral surface could be minimal in this scenario, and indeed Rodriguez-Garcia et al. (2015) have obtained rather long oligopeptides by submitting amino acids solutions to WD cycles in glass vials, without any mineral phase. However, the idea that "surfaces are not necessary after all" would be wrong on two accounts:

– It is unlikely that biomolecules in a realistic prebiotic environment could be highly concentrated enough to precipitate as bulk phases; most probably, the desiccation step would result in dispersion over a mineral surface instead.

− Even in the absence of any significant thermodynamic effect on a reaction, interaction with the surface may greatly modify the reaction kinetics. This may result in reaction inhibition, or conversely, an exaltation in reactivity, that is, catalysis. Catalysis by mineral surfaces, that is, heterogeneous catalysis, is treated in the following paragraph. To stick with the example of peptide bond formation from glycine, for comparable activation conditions, the temperature giving the maximum reaction rate is 80°C lower for glycine on silica than for bulk glycine. This results in a much larger range of stability of peptides formed in mineral surface scenarios compared to bulk scenarios; in the latter case, peptides may decompose almost as soon as they are formed.

9.4. Minerals and reaction kinetics: heterogeneous catalysis

9.4.1. *Lessons from industrial heterogeneous catalysis*

One key feature of geochemical protometabolic scenarios is that enzymatic catalysts have to be replaced by mineral catalysts because efficient enzymes are the product of a long evolutionary history, and even rudimentary enzymes were not present before the first metabolic pathways were established, including those that led to the emergence of proteins or at least rather long peptides. Mineral catalysts have been used in oil refining and the chemical industry for many decades. A huge amount of literature has been devoted to their improvement, and vibrant research is supported by a large scientific community. The selection and use of mineral catalysts in prebiotic chemistry research would definitely benefit from a familiarity with this domain. Of course, not all of industrial catalysis is relevant for the origins of life. In particular, industrials often carry out reactions at high temperatures because increasing the temperature is a simple way to accelerate reactions. This entails a preference for work at the solid/gas interface, rather than at the solid/solution interface. These conditions cannot be considered realistic for the emergence of biochemistry, or only in marginal scenarios. However, interest in reactions carried out in less severe reactions, especially using water as a solvent (Pirkanniemi et al. 2002), is currently growing in the field of industrial catalysis because of the demands of green chemistry. Other current research tendencies in catalytic research include the use of bio-sourced raw materials (Rinaldi et al. 2009) and focus on one-pot reaction sequences. They are interesting with respect to the application to origins of life research, because "hands-off" chemical transformations are obviously desirable.

From a methodological point of view, researchers in industrial catalysis have long developed extensive analysis of reaction kinetics at the interfaces, as well as spectroscopic methods for identifying chemical events and intermediates, which can be valuable for all studies making use of heterogeneous catalysis. The necessity of integrating results from heterogeneous catalysis as well as from geochemistry in origins of life research has been clearly formulated in some research articles (Monnard 2016; Li et al. 2018), but it has not yet been systematically exploited.

9.4.2. What can heterogeneous catalysts do?

The short answer to this question is: a lot. Roughly speaking, industrial catalysts used for large-scale industrial processes can be classified into acido-basic and redox catalysts.

Early Brönsted acidic catalysts employed in the oil industry included clay minerals and other silicates, and starting in the 1960s, zeolites took an overwhelming importance in this field of application. Clays are a mainstay of origins of life research, and zeolites open the way to steric control of reactions, with cages in their structure acting as nanoreactors. Comparison of zeolites with enzymes is very common in the literature, either in general terms or in studies that aim to specifically reproduce the molecular environment provided by the catalytic sites of enzymes in a mineral context (Zecchina et al. 2007; Snyder et al. 2018). Lewis acidic catalysts are also largely used, including alumina and titania. They are active, inter alia, in the catalytic formation of amide bonds (Ali et al. 2015), and thus of peptides from amino acids.

The toolbox of heterogeneous catalysis also includes basic catalysts, among which MgO (and $MgCO_3$) have drawn a lot of attention (Hattori 1995). MgO could have played a role in HCN-based chemistry, often invoked in prebiotic syntheses (Santalucia et al. 2022).

Heterogeneous redox catalysts used industrially for selective oxidations are generally based on transition metal oxides. They are efficient because many transition metals can exist in several oxidation states, and thus temporarily store electrons provided by the reductant and then yield them to the oxidant (or the other way around), allowing a straightforward catalytic mechanism. The redox catalytic properties of most transition elements have been systematically explored. Some of them are rather rare, but research on redox catalysts also focuses on the geochemically significant iron oxides (Sudarsanam et al. 2018), and manganese-containing silicates, or some titanium minerals, may also be of interest in this respect.

An equally large corpus of results has been obtained for redox reactions (e.g. hydrogenation by H_2) on reduced transition metals nanoparticles. While the latter are not supposed to have been commonplace, native copper, nickel, platinum, and silver are mentioned in the list of prebiotic mineral species established by Hazen (2013).

CO_2 fixation in mineral environments is a common preoccupation of prebiotic chemistry (as already mentioned in the Introduction; Wächtershäuser 1992; Muchowska et al. 2017) and industrial chemistry. Not considering photosynthesis, microorganisms can fix CO_2 using five different pathways (Fuchs 2010), and some of them have similarities with directions of research currently tested in CO_2 valorization (Ruiz et al. 2021).

Among the reactions that increase molecular complexity, Fischer-Tropsch is also often invoked as a pathway to complex carbon-containing molecules (Nuth et al. 2008). It consists of the formation of hydrocarbons with a large distribution of molecular weights from "syngas", a mixture of CO and H_2. It was originally developed for the obtention of synthetic gasoline from coal and is still enthusiastically studied for that purpose (van der Laan et al. 1999). Fischer-Tropsch chemistry is quite versatile and can also produce lipids (McCollom Laan et al. 1999) and some nucleobases when NH_3 is added to the mix (Pearce and Pudritz 2005).

Finally, heterogeneous catalysis may be extended to using non-chemical energy sources, in photocatalysis and electrocatalysis. The photocatalytic conversion of CO_2 is often called "artificial photosynthesis" (Tu et al. 2014) and may shed some light on the origins of biological photosynthesis. Electrocatalysis is involved in Russell and Hall's (2002) model for the origin of life in deep submarine vents. Each one of these topics would deserve full coverage, which is not possible here.

9.4.3. Reaction selectivity

Adsorption selectivity has been mentioned in section 9.2.1 as a possible naturally occurring step for the selection of specific molecules from a prebiotic mixture. Many prebiotic scenarios indeed result in a "combinatorial explosion" of related molecules (Colón-Santos et al. 2019), which has given rise to the popular expression "primordial soup" and conjures ideas of intractable complexity. Adsorption selectivity would constitute a post factum pathway to a simplification of prebiotic complexity, but it would be even better to prevent it from arising in the first place. In the cell, enzymes are the catalysts of choice and show a twofold selectivity, for the substrate to be transformed and for the type of chemical reaction to be carried out.

Can heterogeneous catalysts exhibit significant reaction selectivity, even without emulating the performances of enzymes? The state of the art in industrial catalysis suggests that high reaction selectivities can be reached in many systems that have been the object of optimization research for long enough. For example, the selective oxidation of hydrocarbons such as propene by dioxygen can be directed to one among many different products (acrolein, acrylic acid, acetone, benzene, hexadiene, etc.) by choosing the right heterogeneous catalyst, often with selectivities of 98%–99%. Regarding substrate selectivity, there has been less research in the field of industrial catalysis (Lindbäck and Dawaigher 2014) because it is usually more economical to carry out reagent separation prior to applying the catalytic step. However, there exist documented examples using heterogeneous catalysis (Srirambalaji et al. 2012), and Orgel (2008) was certainly right in stating that "discrimination (between very similar substrates) is rare, but not impossible, in reactions catalyzed by small molecules or mineral surfaces".

9.5. A case study: primordial synthesis of pyrimidines

We will illustrate how evaluating the likelihood of extant metabolic pathways in a geochemical context may work in practice from a recent example from our own research (Ter-Ovanessian et al. 2021, 2022, 2023). We recently attempted to reproduce the initial steps of de novo uracil synthesis in a purely mineral setting, corresponding to a fluctuating surface hot spring environment (Damer and Deamer 2004). In the cell, this synthesis involves the so-called "orotate pathway" (Lane and Fan 2015). Its first step consists of the production of an activated molecule, carbamoyl phosphate (CP), from (hydrogeno)carbonate and glutamine, while expending two ATP molecules. It is followed by its reaction with aspartate yielding N-carbamoylaspartate (NCA). Finally, NCA is cyclized to give dihydroorotate (DHO), later to be oxidized to orotate. The main energy input occurs in the first step; each step is kinetically controlled by a specialized enzyme.

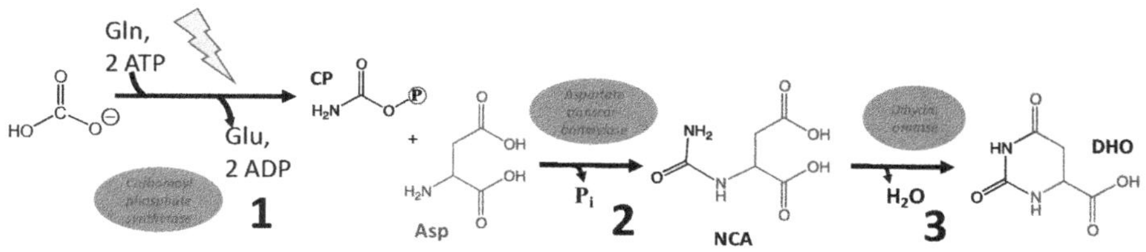

Figure 9.4. *The first steps in de novo pyrimidines synthesis through the orotate pathway. For a color version of this figure, see www.iste.co.uk/dimauro/firststeps.zip*

We first wondered if the biological carbamoylating agent, CP, was likely to have taken part in a prebiotic process. We found that CP is transformed to cyanate in neutral aqueous solutions within a couple of hours at room temperature; in ammoniacal environments, it degrades even faster, within a few minutes. Cyanate is still energy-rich and further transforms to carbamate and then carbonate, but at a much slower rate – significant amounts remain present in solution for several days. A second question was whether CP could be formed from more obvious prebiotic molecules. Trimetaphosphate or P3m is known as an efficient activating agent (see section 9.2.3.1), and its reaction with several carbon-containing small molecules such as urea or carbamate did result in energy transduction, but CP was not part of the observed reaction networks. Thus, we concluded that CP was not a likely participant in prebiotic carbamoylation: there is no obvious pathway for its formation, and if it was present in the solution, it would most likely degrade too quickly to take part in further useful transformations. Useful prebiotic precursors should have a high free energy (or rather free enthalpy of formation), but at the same time they should not be too labile, in case they degrade too quickly to be of any use.

Other potential carbamoylating agents such as urea, cyanate or biuret do not suffer from the same drawbacks as CP, and we investigated their potential reaction with aspartate. A theoretical calculation led us to expect that if equilibrium was reached, aspartate carbamoylation should be extensive with cyanate and even more with biuret, and show minority but still significant yields with urea. In fact, when cyanate was added to aspartate in aqueous solutions (in 1:1 mixtures), NCA was obtained with >90% yield within a few hours at 25°C, without the need for any catalyst. NCA, although still an activated molecule, was very inert as it did not degrade significantly even after one year in solution. In contrast to cyanate, urea and biuret did not give rise to any measurable NCA formation from aspartate, even after several months of reaction – these two molecules are useless as carbamoylating agents due to their inertness. We then checked if the reaction of aspartate with cyanate could operate on a mineral surface as a platform. This was indeed possible on surfaces with basic properties such as $MgCO_3$. Carbamoylation could also occur on silica, since it is a condensation reaction, and silica surfaces are known to efficiently promote condensations (between amino acids, or nucleotides monophosphates) through acid catalysis. However, on silica, cyanate was decomposed to CO_2 + NH_3 before it could react with aspartate. This is due to the fact that every compound that catalyzes a given reaction will also catalyze the reverse one – catalysts for condensation are also catalysts for hydrolysis. Silica became catalytically active at temperatures too low to ensure complete drying so that hydrolysis was favored over condensation.

We then moved on to the next step of the orotate pathway, cyclization of NCA to DHO, a six-membered cycle. This is formally an intramolecular condensation. Heating NCA deposited on silica resulted in cyclization – but to give the five-membered cycle, carboxymethylhydantoin, without any trace of DHO. Catalytic selectivity was indeed observed, but it worked *against* the biological product: the heterogeneous mineral catalyst did not work in the same way as the current enzymatic catalyst. However, different reactivities developed on different surfaces. On $MgCO_3$, NCA remains stable until rather high temperatures where it gets hydrolyzed, perhaps through the participation of lattice hydration water. On zeolites (Na-faujasite), it seems that separate NCA populations follow different fates, perhaps due to their interaction with distinct catalytic sites: the majority undergo a kind of dismutation, where one NCA molecule is hydrolyzed while a second one is condensed to the hydantoin, while a few percent are indeed cyclized to DHO. More research is needed to determine if the fraction transformed to DHO can be maximized by a careful choice of the surface and/or the reaction conditions, or if the hydantoin can branch back into the orotate cycle.

Two lessons may be drawn from this case study. First, different mineral surfaces can promote very different reactions of the same precursors, and we should be wary about the overgeneralization of the potential of mineral surfaces for a given prebiotic reaction, or set of reactions. Second, it is a valid research strategy to investigate whether extant metabolic steps can be duplicated in a geochemical context. Sometimes, the answer is negative: CP can hardly have participated in primordial pyrimidine synthesis. It is conceivable that the pathway to pyrimidine nucleobases started with cyanate as a carbamoylating agent, and that later, when the cyanate supply was exhausted or its prebiotic formation petered out, the reaction network evolved or integrated an alternative network, synthesizing a functionally equivalent molecule, carbamoyl phosphate. The idea that an original nutrient taken from the primordial soup had to be substituted by a self-made metabolite was already formulated by Horowitz (1945). In other cases, the answer is positive: cyanate can very well have carbamoylated aspartate to NCA. And sometimes, the answer is more complex, as for NCA ring closure, but it opens up new directions of research.

9.6. Conclusion

From the preceding developments, it should be clear that mineral surfaces may certainly have played an important role in the emergence of the chemical reactions used by life, but also that a lot remains to be done in order to gain a predictive knowledge of that role. Minerals are not uncommonly encountered in the literature on the origins of life; but often, they are either invoked in untested speculative scenarios, or conversely used in a purely empirical way.

This is partly due to a lack of integration of the methods and concepts of surface science in prebiotic studies. Fortunately, several leading teams are now making full use of the lessons of geochemistry (Hazen and Sverjensky 2010), but the same cannot yet be said for the equally relevant data from industrial heterogeneous catalysis. We believe that many opportunities exist to use the rich chemistry on mineral surfaces in order to advance origins of life studies. These future studies will certainly need to consider reaction thermodynamics and kinetics in precise terms. While some researchers have advocated a sound thermochemical approach of prebiotic reactions (Amend et al. 2013; Barge et al. 2017), these dimensions are still missing from many studies, and we have often been surprised by the lack of fundamental data on even simple biomolecules. Furthermore, life is in fact an open system far from equilibrium exchanging energy and matter with its geochemical environment, and new approaches such as non-equilibrium thermodynamics will probably need to be applied to understand its emergence.

Finally, two important points have hardly been mentioned in this short review, but will no doubt take an important place in future works. The first one is the "vertical" integration of successive protometabolic steps in a coherent scenario with minimal intervention of the experimentalists – such as chemistry under flow in spatially inhomogeneous systems, mimicking geological transport phenomena. The second one is the need to work in realistic environments that encompass a great deal of complexity – "messy environments" to quote the phrase used by Deamer (2022). Only then can significant progress be made toward a robust understanding of the emergence of the complexity of life itself.

9.7. References

Abadian, H., Cornette, P., Costa, D., Mezzetti, A., Gervais, C., Lambert, J.-F. (2022). Leucine on silica: A combined experimental and modelling study of a system relevant for origins of life, and the role of water coadsorption. *Langmuir*, 38(26), 8038–8053.

Ali, M.A., Siddiki, S.M.A.H., Onodera, W., Kon, K., Shimizu, K. (2015). Amidation of carboxylic acids with amines by Nb_2O_5 as a reusable lewis acid catalyst. *ChemCatChem*, 7, 3555–3561.

Amend, J.P., LaRowe, D.E., McCollom, T.M., Shock, E.L. (2013). The energetics of organic synthesis inside and outside the cell. *Phil. Trans. R. Soc. B*, 368, 20120255.

Arrhenius, G., Sales, B., Mojzsis, S., Lee, T. (1997). Entropy and change in molecular evolution – The case of phosphate. *J. Theor. Biol.*, 187, 503–522.

Barge, L.M., Branscomb, E., Brucato, J.R., Cardoso, S.S.S., Cartwright, J.H.E., Danielache, S.O., Galante, D., Kee, T.P., Miguel, Y., Mojzsis, S., et al. (2017). Thermodynamics, disequilibrium, evolution: Far-from-equilibrium geological and chemical considerations for origin-of-life research. *Orig. Life Evol. Biosph.*, 47, 39–56.

Bernal, J.D. (1949). The physical basis of life. *Proc. Phys. Soc. London (Sect. A)*, 62(357), 537–558.

Boujday, S., Lambert, J.-F., Che, M. (2003). Evidence for interfacial molecular recognition in transition metal complexes adsorption on amorphous silica surfaces. *J. Phys. Chem. B*, 107(3), 651–654.

Castner, D.G. and Ratner, B.D. (2002). Biomedical surface science: Foundations to frontiers. *Surf. Sci.*, 500, 28–60.

Colón-Santos, S., Cooper, G.J.T., Cronin, L. (2019). Taming the combinatorial explosion of the formose reaction via recursion within mineral environments. *ChemSystemsChem*, 1, e1900014.

Costanzo, G., Saladino, R., Crestini, C., Ciciriello, F., Di Mauro, E. (2007). Nucleoside phosphorylation by phosphate minerals. *J. Biol. Chem.*, 282(23), 16729–16735.

Damer, B. and Deamer, D. (2004). The hot spring hypothesis for an origin of life. *Astrobiol.*, 20(4), 429–452.

Deamer, D. (2022). Origins of life research: The conundrum between laboratory and field simulations of messy environments. *Life (Basel)*, 12, 1429.

de Duve, C. and Miller, S.L. (1991). Two-dimensional life? *Proc. Natl. Acad. Sci. USA*, 88, 10014–10017.

El Samrout, O., Fabbiani, M., Berlier, G., Lambert, J.-F., Martra, G. (2022). Emergence of order in origin-of-life scenarios on mineral surfaces: Polyglycine chains on silica. *Langmuir*, 38, 15516–15525.

Fabbiani, M., Pazzi, M., Vincenti, M., Tabacchi, G., Fois, E., Martra, G. (2018). Does the abiotic formation of oligopeptides on TiO_2 nanoparticles require special catalytic sites? Apparently not. *J. Nanosc. Nanotechnol.*, 18(8), 5854–5857.

Ferris, J.P., Hill, A.R.J., Liu, R., Orgel, L.E. (1996). Synthesis of long prebiotic oligomers on mineral surfaces. *Nature*, 381, 59–61.

Fuchs, G. (2010). Alternative pathways of carbon dioxide fixation: Insights into the early evolution of life? *Ann. Rev. Microbiol.*, 65(1), 631–658.

Gibard, C., Bhowmik, S., Karki, M., Kim, E.-K., Krishnamurthy, R. (2018). Phosphorylation, oligomerization and self-assembly in water under potential prebiotic conditions. *Nature Chem.*, 10, 212–218.

Gorlero, M., Wieczorek, R., Adamala, K., Giorgi, A., Schininà, M.E., Stano, P., Luisi, P.L. (2009). Ser-His catalyses the formation of peptides and PNAs. *FEBS Lett.*, 58(3), 153–156.

Greenwell, H.C. and Coveney, P.V. (2006). Layered double hydroxide minerals as possible prebiotic information storage and transfer compounds. *Orig. Life Evol. Biosph.*, 36, 13–37.

Hassenkam, T., Damer, B., Mednick, G., Deamer, D. (2020). AFM images of viroid-sized rings that self-assemble from mononucleotides through wet–dry cycling: Implications for the origin of life. *Life (Basel)*, 10(321).

Hattori, H. (1995). Heterogeneous basic catalysis. *Chem. Rev.*, 95, 537–558.

Hazen, R.M. (2013). Paleomineralogy of the Hadean eon: A preliminary species list. *Am. J. Sci.*, 313, 807–843.

Hazen, R.M. and Sverjensky, D.A. (2010). Mineral surfaces, geochemical complexities, and the origins of life. *Cold Spring Harb. Perspect. Biol.*, 2, a002162.

Hazen, R.M., Filley, T.R., Goodfriend, G.A. (2001). Selective adsorption of L- and D-amino acids on calcite: Implications for biochemical homochirality. *Proc. Natl. Acad. Sci. USA*, 98(10), 5487–5490.

Higgs, P.G. (2016). The effect of limited diffusion and wet–dry cycling on reversible polymerization reactions: Implications for prebiotic synthesis of nucleic acids. *Life (Basel)*, 6, 24.

Horowitz, N.H. (1945). On the evolution of biochemical syntheses. *Proc. Nat. Acad. Sci. USA*, 31(6), 153–157.

Huber, C., Eisenreich, W., Hecht, S., Wächtershäuser, G. (2003). A possible primordial peptide cycle. *Science*, 301, 938–940.

Ianeselli, A., Atienza, M., Kudella, P.W., Gerland, U., Mast, C.B., Braun, D. (2022). Water cycles in a Hadean CO_2 atmosphere drive the evolution of long DNA. *Nature Phys.*, 18, 579–585.

Jaber, M., Georgelin, T., Bazzi, H., Costa-Torro, F., Lambert, J.-F., Bolbach, G., Clodic, G. (2014). Selectivities in adsorption and peptidic condensation in the (arginine and glutamic acid)/montmorillonite clay system. *J. Phys. Chem. C*, 118, 25447–25455.

Keller, M., Blochl, E., Wächtershäuser, G., Stetter, K.O. (1994). Formation of amide bonds without a condensation agent and implications for origin of life. *Nature*, 368, 836–838.

Kloprogge, J.T.T. and Hartman, H. (2022). Clays and the origin of life: The experiments. *Life (Basel)*, 12, 259.

van der Laan, G.P. and Beenackers, A.A.C.M. (1999). Kinetics and selectivity of the Fischer-Tropsch synthesis: A literature review. *Catal. Rev. – Sci. Eng.*, 41(3–4), 255–318.

Lahav, N. and Chang, S. (1976). The possible role of solid surface area in condensation reactions during chemical evolution: Reevaluation. *J. Mol. Evol.*, 8, 357–380.

Lane, A.N. and Fan, T.W.-M. (2015). Regulation of mammalian nucleotide metabolism and biosynthesis. *Nucl. Acids Res.*, 43(4), 2466–2485.

Lazcano, A. and Miller, S.L. (1999). On the origin of metabolic pathways. *J. Mol. Evol.*, 49, 424–431.

Leman, L., Orgel, L., Ghadiri, M.R. (2004). Carbonyl sulfide–mediated prebiotic formation of peptides. *Science*, 306, 283–386.

Li, Y.M., Kitadai, N., Nakamura, R. (2018). Chemical diversity of metal sulfide minerals and its implications for prebiotic catalysis. *Life*, 8(4), 46.

Lindbäck, E., Dawaigher, S., Wärnmark, K. (2014). Substrate-selective catalysis. *Chem. Eur. J.*, 20, 13432–13481.

Marshall-Bowman, K., Ohara, S., Sverjensky, D.A., Hazen, R.M., Cleaves, H.J. (2010). Catalytic peptide hydrolysis by mineral surface: Implications for prebiotic chemistry. *Geochim. Cosmochim. Acta*, 74(20), 5852–5861.

Martin, W.F. (2019). Carbon–metal bonds: Rare and primordial in metabolism. *Trends Biochem. Sci.*, 44(9), 807–818.

Martin, W.F. and Russell, M.J. (2007). On the origin of biochemistry at an alkaline hydrothermal vent. *Phil. Trans. R. Soc. B*, 362, 1887–1925.

Maurel, M.-C. and Décout, J.-L. (1999). Origins of life: Molecular foundations and new approaches. *Tetrahedron*, 55, 3141–3182.

McCollom, T.M., Ritter, G., Simoneit, B.R.T. (1999). Lipid synthesis under hydrothermal conditions by Fischer-Tropsch-type reactions. *Orig. Life Evol. Biosph.*, 29(2), 153–166.

Monnard, P.-A. (2016). Taming prebiotic chemistry: The role of heterogeneous and interfacial catalysis in the emergence of a prebiotic catalytic/information polymer system. *Life*, 6, 40.

Muchowska, K.B., Varma, S.J., Chevallot-Beroux, E., Lethuillier-Karl, L., Li, G., Moran, J. (2017). Metals promote sequences of the reverse Krebs cycle. *Nature Ecol. Evol.*, 1(11), 1716–1721.

Muchowska, K.B., Varma, S.J., Moran, J. (2020). Nonenzymatic metabolic reactions and life's origins. *Chem. Rev.*, 120, 7708–7744.

Nuth, J.A., Johnson, N.M., Manning, S. (2008). A self-perpetuating catalyst for the production of complex organic molecules in protostellar nebulae. *Astrophys. J.*, 673, L225–L228.

Orgel, L.E. (2000). Self-organizing biochemical cycles. *Proc. Nat. Acad. Sci. USA*, 97(23), 12503–12507.

Orgel, L.E. (2006). In the beginning. *Nature*, 439, 915.

Orgel, L.E. (2008). The implausibility of metabolic cycles on the prebiotic earth. *PLoS One*, 6(1), e18.

Osterberg, R. and Orgel, L.E. (1972). Polyphosphate and trimetaphosphate formation under potentially prebiotic conditions. *J. Mol. Evol.*, 1 (241–248).

Patel, B.H., Percivalle, C., Ritson, D.J., Duffy, C.D., Sutherland, J.D. (2015). Common origins of RNA, protein and lipid precursors in a cyanosulfidic protometabolism. *Nature Chem.*, 7(4), 301–307.

Pearce, B.K.D. and Pudritz, R.E. (2015). Seeding the pregenetic earth: Meteoritic abundances of nucleobases and potential reaction pathways. *Astrophys. J.*, 807(1), 85.

Pirkanniemi, K. and Sillanpaa, M. (2002). Heterogeneous water phase catalysis as an environmental application: A review. *Chemosphere*, 48(10), 1047–1060.

Preiner, M., Igarashi, K., Muchowska, K.B., Yu, M.Q., Varma, S.J., Kleinermanns, K., Nobu, M.K., Kamagata, Y., Tüysüz, H., Moran, J. et al. (2019). A hydrogen dependent geochemical analogue of primordial carbon and energy metabolism. *Nature Ecol. Evol.*, 4(4), 534–542.

Ralser, M. (2018). An appeal to magic? The discovery of a non-enzymatic metabolism and its role in the origins of life. *Biochem. J.*, 475, 2577–2592.

Rimola, A., Costa, D., Sodupe, M., Lambert, J.-F., Ugliengo, P. (2013). Silica surface features and their role in the adsorption of bio-molecules: Computational modeling and experiments. *Chem. Rev.*, 113, 4216–4313.

Rinaldi, R. and Schüth, F. (2009). Design of solid catalysts for the conversion of biomass. *Energ. Environ. Sci.*, 2(6), 610–626.

Rodriguez-Garcia, M., Surman, A.J., Cooper, G.J.T., Suárez-Marina, I., Hosni, Z., Lee, M.P., Cronin, L. (2015). Formation of oligopeptides in high yield under simple programmable conditions. *Nature Com.*, 6, 8385.

Ross, D.S. and Deamer, D. (2016). Dry/wet cycling and the thermodynamics and kinetics of prebiotic polymer synthesis. *Life*, 6(28), Article E28.

Ruiz, P., Fernández, C., Ifandi, E., Eloy, P., Meza-Trujillo, I., Devred, F., Gaigneaux, E.M., Tsikouras, B. (2021). Abiotic transformation of H_2 and CO_2 into methane on a natural chromitite rock. *ACS Earth Space Chem.*, 5, 1695–1708.

Russell, M.J. and Hall, A.J. (2002). From geochemistry to biochemistry. Chemiosmotic coupling and transition element clusters in the onset of life and photosynthesis. *The Geochem. News*, 113, 6–12.

Santalucia, R., Pazzi, M., Bonino, F., Signorile, M., Scarano, D., Ugliengo, P., Spoto, G., Mino, L. (2022). From gaseous HCN to nucleobases at the cosmic silicate dust surface: An experimental insight into the onset of prebiotic chemistry in space. *Physical Chemistry Chemical Physics*, 24(12), 7224–7230.

Shock, E.L. (1992). Stability of peptides in high-temperature aqueous solutions. *Geochim. Cosmochim. Acta*, 56, 3481–3491.

Slowing, I.I., Trewyn, B.G., Giri, S., Lin, V.S.-Y. (2007). Mesoporous silica nanoparticles for drug delivery and biosensing applications. *Adv. Funct. Mater.*, 17(8), 1225–1236.

Smith, J.V. (1998). Biochemical evolution I: Polymerization on internal, organophilic silica surfaces of dealuminated zeolites and feldspars. *Proc. Natl. Acad. Sci. USA*, 95, 3370–3375.

Snyder, B.E.R., Bols, M.L., Schoonheydt, R.A., Sels, B.F., Solomon, E.I. (2018). Iron and copper active sites in zeolites and their correlation to metalloenzymes. *Chem. Rev.*, 118(5), 2718–2768.

Srirambalaji, R., Hong, S., Natarajan, R., Yoon, M., Hota, R., Kim, Y., Ko, Y.H., Kim, K. (2012). Tandem catalysis with a bifunctional site-isolated Lewis acid-Brønsted base metal-organic framework, NH_2-MIL-101(Al). *Chem. Commun.*, 48, 11650–11652.

Sudarsanam, P., Zhong, R.Y., Van den Bosch, S., Coman, S.M., Parvulescu, V.I., Sels, B.F. (2018). Functionalised heterogeneous catalysts for sustainable biomass valorisation. *Chem. Rev.*, 47, 8349.

Ter-Ovanessian, L., Rigaud, B., Mezzetti, A., Lambert, J.-F., Maurel, M.-C. (2021). Carbamoyl phosphate and its substitutes for the uracil synthesis in origins of life scenarios. *Sci. Rep.*, 11, 19356.

Ter-Ovanessian, L., Lambert, J.-F., Maurel, M.-C. (2022). Building the uracil skeleton in primitive ponds at the origins of life: Carbamoylation of aspartic acid. *Scientific Reports*, 12, 19178.

Ter-Ovanessian, L.M.P., Lambert, J.-F., Maurel, M.-C. (2023). Ring-closures on the rocks in a prebiotic environment. *ChemBioChem*, 24, e202300143.

Tu, W.G., Zhou, Y., Zou, Z.G. (2014). Photocatalytic conversion of CO_2 into renewable hydrocarbon fuels: State-of-the-art accomplishment, challenges, and prospects. *Adv. Mater.*, 26, 4607–4626.

Wächtershäuser, G. (1988a). Before enzymes and templates: Theory of surface metabolism. *Microbiol. Rev.*, 452.

Wächtershäuser, G. (1988b). Pyrite formation, the first energy source for life: A hypothesis. *System. Appl. Microbiol.*, 10, 207–210.

Wächtershäuser, G. (1992). Groundworks for an evolutionary biochemistry: The iron-sulfur world. *Progr. Biophys. Molec. Biol.*, 58, 85–201.

Yamagata, Y., Watanabe, H., Saitoh, M., Namba, T. (1991). Volcanic production of polyphosphates and its relevance to prebiotic evolution. *Nature*, 352(6335), 516–519.

Zecchina, A., Rivallan, M., Berlier, G., Lambert, C., Ricchiardi, G. (2007). Structure and nuclearity of active sites in Fe-zeolites: Comparison with iron sites in enzymes and homogeneous catalysts. *Phys. Chem. Chem. Phys.*, 9, 3483–3499.

10

A Rationale for the Evolution of the Genetic Code in Relation to the Stability of RNA and Protein Structures

Andrew TRAVERS
MRC Laboratory of Molecular Biology, Cambridge Biomedical Campus, University of Cambridge, United Kingdom

Understanding the evolution of the genetic code is integral to our perception of the origin of life itself and has long been a matter for discussion and debate. I argue here that the initial selection of codon–anticodon pairs depended on the conservation of a relatively stable stem-loop structure containing the anticodon and also the stabilization of single-stranded mRNA structures by base-stacking. The structures adopted by single-stranded RNAs have been shown to be base dependent and would facilitate the formation of a "hemi" kissing loop to define the codon–anticodon interaction. This scenario requires that the evolution of structural stability was a necessary precursor to the establishment of the genetic code. Similarly, the evolution of the structural stability of proteins would have similar requirements and promote the precision necessary for establishing rapid reaction rates underlying Darwinian evolution. This process would possibly be concomitant with the late accession of appropriate amino acids to the code.

10.1. Introduction

The temporal evolution of the genetic code has long been a topic of controversial discussion, yet the origins of the present-day code are crucial to our understanding of not only the code itself, but also to the evolution of proteins and how they are able to do the work necessary to maintain biological homeostasis. Discussions initially centered on the triplet nature of the code (Crick 1968, 1976), and then on the relative thermal stability of the necessary codon–anticodon interactions in an RNA world (Trifonov 2000; Travers 2006). Subsequently, evidence was adduced from the presumed ancient conservation of sequence motifs, particularly those associated with the functions of the adenine and guanine nucleotides (Trifonov 2009). Recently, attention has been directed both to the changing and assumed evolutionary directed patterns of utilized amino acids in whole genomes (Jordan 2005) and also to an assumed selection for optimization of error minimization in the translation of mRNA coding sequences (Koonin 2017).

These aspects of the evolution of the code all, to a greater or less extent, have validity, but the question remains: is a specific characteristic of the biological chemical system being selected for – and, if so, which – or, more probable, an amalgam of several different important characteristics? At the core of this question is the notion of the nature of biological complexity. This can be regarded as a dynamic mixture of regulated coordinated chemical reactions (Travers 2022), evolved for each organism to be effective in a particular biological niche. On this view, evolution of both the codon–anticodon interaction and also of protein function, and hence of the mix of utilized amino acids, depends on relative structural stability, which facilitates the overall efficiency of the chemical processes with respect to the cumulative rates of chemical reactions involved. Darwinian selection, and hence evolution, would likely act primarily on this parameter and is consequently related to the efficiency of catalysis and hence to the mechanism of protein function determined by the associated structures.

10.2. Codon–anticodon recognition

A primary question in considerations of the evolution of the genetic code is how both the accuracy and stability of the crucial codon–anticodon recognition step can be ensured. The anti-codon is presented to an incoming codon within the apex of a short stem-loop structure (reviewed in Grosjean (2016)). Such a structure should be conformationally constrained relative to an otherwise unstructured RNA strand and consequently, is likely a major determinant of the geometry of the codon–anticodon minihelix formed primarily by base pairing. The stem itself contains a highly conserved G-C or C-G Watson–Crick base pair above the loop, and immediately above the loop is a base-pair whose character is reflected in the organization of the genetic code into boxes (Grosjean 2016). This latter base-pair is a component of the "extended" anti-codon proposed by Yarus (1982). In the ribosome, the short stem-loop structure makes further highly conserved contacts with residues A1492 and A1493 of helix 44 of 16S RNA (Ogle 2005; Demeshkina 2012). It has been hypothesized that it is this presentation of the anti-codon sequence in the ribosome acceptor-site that optimizes the formation of Watson–Crick base-pairs with the first two bases of the mRNA codon. In other words, the stem imparts a relative rigidity to the structure of the anti-codon loop.

A similar argument has been suggested for the presentation of the codon within a short, stabilized loop. Indeed, tRNAs with complementary anticodons readily form anticodon–anticodon base-pairs (Eisinger 1971; Grosjean 1976, 1978) – a short kissing-loop interaction – but any reliance on stabilized loops would ultimately not be compatible with the current efficient continuous triplet readout. In this scenario, any sequence-dependent effect that stabilized the structure of a free mRNA strand could have the potential to facilitate codon–anticodon base-pairing. Indeed, recent experiments (Seol 2007; Plumridge 2020) on the structure of homopolymeric RNA strands have been interpreted to show that these polymers can adopt preferred conformations whose nature depends on the base. Thus, Seol et al. (2007) provided evidence from single molecule stretching experiments that both poly(rA) and poly(rC) could undergo a coil-helix transition to form a helix, with that formed by poly(rC) being more stable than that formed by poly(rA). Similarly, Plumridge et al. (2020) compared poly(rA) and poly(rU) applying a number of different experimental techniques. They found that chains of rU bases are relatively unstructured under all conditions, while chains of rA bases were more ordered with a high degree of base-stacking, the extent of which was dependent on the ionic environment, especially the presence of Mg^{2+}. Although, possibly for technical reasons, there are no comparable data for poly(rG), the stability of secondary structure formation by RNA homopolymers has the tentative order poly(rC)>poly(rA)>poly(rU) with that of poly(rG) undetermined.

The ability of RNA homopolymers to form structured single stranded helices in which the presentation of successive bases to the solvent, and potentially to complementary bases, is relatively uniform suggests a simple model for primeval codon–anticodon recognition. A structured anticodon loop would base-pair with a single-stranded RNA helix to form, in effect, a hemi kissing structure (stem-loop interacting with a structured mRNA). The formation of kissing-loops of course depends on the strength of the participating base-pairs, but also strongly on stabilizing base-stacking interactions (Bouchard 2014; Schulz 2017). In the context of the genetic code, there are no comprehensive comparative data relating to the relative stability of kissing loops with different sequences. However, one pertinent observation is that a tRNA^{leu} with the anticodon UAG forms an anomalously unstable complex with the complementary anticodon CUA of tRNA$^{tyr(su+)}$ (Grosjean 1978). Except for one base modification, the base composition of this interaction is identical to that of the most stable complex examined, with UUC (tRNA^{phe}) and GAA (tRNA^{glu}) as the complementary triplets. Even after allowing for the effect of the additional base modification, the authors concluded that the UAG-CUA complex was still anomalously weak. They speculated that the sequence UAG may have been selected as a termination codon because its complementary complex is too weak. In other words, termination would have been the default when a codon triplet was not stably engaged with an anticodon.

To account for the apparent organization of the genetic code into four-codon and two-codon boxes, both Jukes (1967) and Crick (1968) hypothesized that the initial code would have been a two-letter triplet of the form XYNXYN... in which the codon–anticodon interactions would have been mediated by the XY base step. A coding box XYN would, on this model, specify a single amino acid. Subsequently, further accessions that allowed discrimination between three-letter triplets utilizing the wobble rules would have developed, generating a code of the form XYZXYZ... . In this case, a coding box could specify two or in some cases even more amino acids.

A consideration of the stabilising effects of base-stacking in both kissing loops and hemi kissing loops bears on the core arguments for ordering the progressive evolutionary accessions of codons – and hence amino acids – to the genetic code. The concept that the primacy of codon acquisition in the development of the genetic code is dependent on the stability of the codon–anticodon complement was first advanced by Eigen and Schuster (1978). This robust concept was further developed by Trifonov (2000, 2004) who argued that the thermostability rule, when applied to coding triplets, was consistent, with only a few exceptions, with the deduced chronological accretion of codons to the standard code. Strikingly, the most ancient amino acids in his derived temporal order mirrored the nine most abundant encoded amino acids in Miller's recreation of the hypothetical primordial environment (Miller 1953; Miller and Urey 1959). This correlation of thermostability with the chronology of code evolution depends critically on the accuracy of the values assigned to the stability of the individual base steps. Recently, the favored experimental method for determining these values for both RNA and DNA was to determine the melting temperature for oligonucleotides of defined length sequence and then deconvolute the parameters for base steps (i.e. dinucleotides) based on base step composition of the tested sequences (SantaLucia 1998; Xia et al. 1998).

Suggestions for the chronology differ in two major respects – the absolute primacy of four-codon boxes and the positioning of the termination codons UGA, UAA and UAG which, on the argument presented above, could effectively constitute the minimum energy associated with a codon–anticodon interaction (Trifonov 2000; Travers 2006; Grosjean 2016). The ordering of Trifonov (2000) and Grosjean (2016) draw on data for the thermal stability of short RNA duplexes (~12–20 bp) (Xia 1998), which however are significantly longer that the mini-helix invoked for the codon–anticodon interaction (3 bp), while that of Travers (2007) uses data derived from the experimental determination of stacking energies at DNA nicks, from which the melting energies were calculated (Protozanova 2004). The two data sets differ in one major respect – the melting energies assigned to the purine-pyrimidine base steps TA (UA), TG/CA (UG/CA) and CG. The values derived from DNA stacking energies are uniformly the lowest (least stable) of base steps of the same base composition, whereas those derived from RNA–RNA annealing all have a relatively higher assigned melting energy. For example, the DNA data place TA as the least stable step by far, whereas the RNA data show that UA/UA is more stable than the AA/UU and AU/AU steps. Although the thymine methyl group influences base stacking, similar, less pronounced differences are also apparent between the melting energies obtained from the DNA stacking data and from DNA annealing data using the same protocol as that for the derivation of the RNA annealing data (SantaLucia 1998). While the origin of these differences is controversial (Vologodskii 2018), their relevance to the relative stabilities of kissing loops and related structures to longer duplexes is crucial to speculations on the origins and the subsequent genetic code.

In the most comprehensive analysis of the ordering of the evolutionary appearance of codons, Trifonov (2000) employed multiple criteria in addition to the computed melting energies for RNA base-pairs (Xia). These criteria were then weighted to correct for apparent redundancy. On this basis, as a working model, Trifonov's derived order for the chronology of codon appearance was **Gly/Ala**, Val/**Asp**, Pro, Ser, **Glu**/Leu, Thr, Arg, Asn, Lys, Gln, Ile, Cys, His, Phe, Met, Tyr and Trp, where the amino acids in bold are those which were detected in Miller's experiments with electric discharge to imitate assumed conditions of early Earth atmosphere (Miller 1953). Of the 20 amino acids, the first 10, with the exception of glutamic and aspartic acids, include all the four-codon boxes. The ordering of the remaining core amino acids was then more dependent on criteria other than the assigned melting energies and Miller observations. Taking into account both the distinction between the early and late codons (Trifonov 2000) and also the observation that four-codon boxes always contain either a C–G codon–anticodon pair at second position or a C–G/G–C pair at first position (Grosjean 2016), the presentation of the genetic code can be rearranged from the classic format (Figure 10.1). What emerges is a pattern in which all the early codons and four-codon boxes are placed in the top left-hand half diagonally opposite all the two-box codons and the later codon accessions. On this basis – and this is also true of alternative presentations – the latest amino acid additions to the code, selenocysteine and pyrrolysine, are located in the bottom right-hand half of the diagram.

The only differences between the organization presented in Figure 10.1 and that proposed by Grosjean (2016) are in the clustering of all the four-codon boxes and the separation of the chronological order into early and late accessions. They are otherwise similar. An implicit assumption in the use of the RNA thermal stability data (Xia 1998) is that the codon–anticodon mini-helix adopts an A-form structure. For the third (wobble) position, this is unlikely to be the case (Murphy IV and Ramakrishnan 2004). Further analysis of the structure of poly(rA) structures in the presence of Mg^{2+} indicates a shorter length than that expected for a purely A-form stacking interaction (Plumridge 2020). If it is assumed that any stacking interactions responsible for maintaining the stability and accuracy of the mini-helix are similar to those in DNA, one role of the stem-loop and anticodon modifications would be to modulate the stacking interactions between neighboring bases by making minor adjustments to their charge distribution (Hunter 1993). Examples would be the methylation of C and U residues. One other possible correlation between codon sequences and stacking is the altered assignment of codons. Including the addition of selenocysteine and pyrrolysine to the code, 80% (20/25) of the different assignments (listed in Koonin (2018)) contain a pyrimidine-purine base-step in the codon sequence. In the DNA analysis arguing from base-stacking, these are among the least stable base-steps. The value of 80% compares with the frequency of 50% expected from random alternative assignment. This observation also includes the adoption of the ATG codon by methionine.

sR		sR		iR		iR	
GCU	ala	GGU	gly	GUU	val	GAU	asp
GCC	ala	GGC	gly	GUC	val	GAC	asp
GCA	ala	GGA	gly	GUA	val	GAA	glu
GCG	ala	GGG	gly	GUG	val	GAG	glu

sL		sL		iL		iL	
CCU	pro	CGU	arg	CUU	leu	CAU	his
CCC	pro	CGC	arg	CUC	leu	CAC	his
CCA	pro	CGA	arg	CUA	leu	CAA	gln
CCG	pro	CGG	arg	CUG	leu	CAG	gln

iL		iL		wL		wL	
ACU	thr	AGU	*ser*	AUU	ileu	AAU	asn
ACC	thr	AGC	*ser*	AUC	ileu	AAC	asn
ACA	thr	AGA	*arg*	AUA	ileu	AAA	lys
ACG	thr	AGG	*arg*	AUG	met	AAG	lys

iR		iR		wR		wR		
UCU	ser	UGU	cys	UUU	phe	UAU	tyr	
UCC	ser	UGC	cys	UUC	phe	UAC	tyr	
UCA	ser	UGA	stop -> Sec	UUA	*leu*	UAA	stop	
UCG	ser	UGG	trp	UUG	*leu*	UAG	stop	->Pyl

Figure 10.1. *For a color version of this figure, see www.iste.co.uk/dimauro/firststeps.zip*

COMMENT ON FIGURE 10.1.– *The first 10 codons of the Trifonov chronology are shown in bold and the final six in red. Those intermediate between these two classes (Gln, Asn, Ileu, Lys) are shown in yellow. Possible later expansions of the coding capacity for Ser, Arg and Leu are given in italic. In the middle two columns, the separation between four- and two-box codons is indicated by a line. The terms sL, sR, iL, iR, wL and wR above the boxes refer to the circular formulation of Grosjean (2016). s, i and w, respectively, indicate strong, intermediate and weak codons, while L and R indicate the left and right sides of the circle. The restriction of L and R to individual lines is an artifact of the sequence organization of the original circle. Sec = selenocysteine and Pyl = pyrrolysine.*

In all analyses, the chronological position of arginine remains an enigma. It occupies one of the four-codon boxes, a position consistent with both sets of thermodynamic parameters used. Nevertheless, not only has it not been identified as a product of electric discharge experiments (Miller 1953; Criado-Reyes 2021), but its current biosynthetic pathways are complex, as in eukaryotes involving the co-operation of two cellular compartments. No probiotic pathway for its synthesis, nor obvious precursors, has yet been identified. However, carbodiimide, a potential source of the arginine guanidinium group, is produced by formamide reactions in the presence of certain minerals (Saladino 2018). Dicyanamide would be another possibility (Travers 2006).

The failure to identify arginine in electric discharge experiments illustrates a more general problem in chronology derivation. Although the amino acids identified by Miller (1953) fit nicely with other criteria (e.g. Trifonov 2000), later experiments indicate that this correlation is not wholly secure because the relative yield as well as the nature of the amino acids generated is variable. For example, Criado-Reyes (2021) observed that in addition to those identified by Miller (1953) (with the notable exception of threonine), lysine, asparagine and histidine, all of which fall into the category of late accessions (Figure 10.1), can be generated by electric discharge. Additionally, other amino acids, which are not included in the extant code, are generated by these procedures.

The selection of the coded amino acids has often been ascribed to "frozen accidents". While there is a case to be made for the initial selections on this basis, a complementary view is that Darwinian selection of the system acts on protein function and this can depend on the nature of the included amino acids.

In modern day proteins, the active center of many enzymes is almost invariably associated with a structurally rigid region at the center of the molecule (Chalopin 2020). A good example is the *E. coli* RNA polymerase, an assembly of approximately 3,500–4,000 amino acids. In this assembly, the central rigid region contains the active site, which in RNA polymerase contains a loop with three closely spaced aspartic residues (the Asp triad) coordinated with a tightly bound Mg^{2+} ion (Steitz 1998; Gangurde 2000). This same distribution of rigidity is also a characteristic of many simpler globular enzymes. Indeed, in one study (Chalopin 2020), over 800 enzymes shared this pattern.

One rationale for this conserved pattern of rigidity is that it enhances the rates of enzyme-catalyzed reactions. In a dynamic multi-component chemical system, it is precisely these rates that are potentially a major basis for Darwinian selection and the further evolution of the system. In this context, we need to understand the structural basis for localized rigidity within enzymes and, indeed, within other proteins, including receptors and regulated ion channels. Relative rigidity in this context is defined as local minima in the pattern of crystallographic temperature factors (B-factors) (Schlessinger 2005; Yuan 2005). The major difference between regions on the outside of a protein or protein assembly where the polypeptide chains are "soft" and the center where they are rigid is, by definition, related to the elastic constant and more specifically likely to the variable torsional constants, C, of the chain (Chalopin 2018). At the center, C is higher and the chain "stiffer". This variation correlates with the amplitude and frequency of high-frequency vibrations, which are maximal at the center (Chalopin 2020). The wavelengths of these motions, and consequently configurational fluctuations of the chain, are correspondingly minimized at this location.

In protein structures, the amino acids classified as late accessions to the code are Cys, His, Phe, Met, Tyr and Trp (Trifonov 2000). With the possible exception of Cys, all these amino acids are often located in hydrophobic pockets where they interact via electron delocalization, for example, the $\pi-\pi$ bonding between aromatic rings and delocalized interactions between a ring and the Me-S of Met, although complex sidechain groups of His, Phe, Tyr and Trp are internally dynamically rigid relative to that of arginine. Cys, in addition to its ability to form S-S bridges, can also contribute to local rigidity by forming coordination complexes with Zn^{2+}. The selection pressures for the initial incorporation of these amino acids into the code are probably unknowable. However, since the amino acid composition of proteins is likely far from total equilibrium (Jordan 2005), subsequent amino acid substitutions can potentially change the relative proportions of early and late accessions on a genome-wide basis. An analysis of the patterns of amino acid composition drift during evolution revealed five strong "gainers" (Cys, Met, His, Ser and Phe) and four strong "losers" (Pro, Ala, Glu and Gly) (Jordan 2005). Apart from Ser, the "gainers" are all late accessions to the code, while the "losers" are among the earliest. Although a structural rationale, if any, for these changes in pattern is unknown, it seems plausible that the drift is related to protein function and/or folding (Komar 2021).

10.3. Concluding remarks

What does the variation in the products of the electric discharge experiments tell us about the origin of the code? Certainly, it indicates the nature of readily synthesized amino acids, but electric discharge is only one suggested route to prebiotic amino acid synthesis. Reactions involving formamide also produce, among many other products, glycine, proline and alanine, especially in the presence of silicate (Saladino 2019). All three of these are considered to be among the earliest of the encode amino acids. Also unlike natural discharges (lightning), as invoked for Darwin's "warm little pond" (Pearce 2017), the supply of energy in alkaline silicaceous environment in the Hadean era is likely to be more constant (Mulkidjanian 2012; Dibrova 2015; Boehnke 2018; Travers 2022). In the context of origins, this review has raised two issues. What is the role, if any, of base stacking in maintaining the stability of both anticodon–codon complexes and structured mRNA? Here, more data are a necessity. And also to what extent did Mg^{2+} or a similar ion participate in further definition of the structures? In the present-day ribosomes, the kink in the mRNA between the A- and P-sites is associated with Mg^{2+} ions (Demeshkina 2015). It may not be coincidental that one of the proposed nurseries for the origin of life implicates the conversion of olivine to serpentine, both magnesium-rich minerals (see Saladino (2019) for a full discussion). Fundamentally, the discussion on the stability of nucleic acid interactions and proteins implies an evolution toward a more organized and structured system, one that would increasingly diverge from thermodynamic equilibrium with the external environment. The evolution of such a system implies not only the almost continuous availability of a source of energy, but also the presence of an initial environment that itself was not in thermodynamic equilibrium, at least transiently. Such a requirement would be satisfied by a naturally occurring energy-maintained electrochemical gradient (Schultz 1961; Travers 2022).

10.4. Acknowledgments

The author acknowledges support by the Medical Research Council MC_U105178783.

10.5. References

Anderson, P. and Bauer, W. (1978). Supercoiling in closed circular DNA: Dependence on ion type and concentration. *Biochemistry*, 17, 594–601.

Boehnke, P., Bell, E.A., Stephan, T., Trappitsch, R., Keller, C.B., Pardo, O.S., Davis, A.M., Harrison, T.M., Pellin, M.J. (2018). Potassic, high silica Hadean crust. *Proc. Natl. Acad. Sci. USA*, 115, 6353–6356.

Bouchard, P. and Legault, P. (2014). A remarkably stable kissing loop interaction defines substrate recognition by the *Neurospora* Varkud satellite enzyme. *RNA*, 20, 1451–1454.

Chalopin, Y. (2020). The physical origin of rate promoting vibrations in enzymes revealed by structural rigidity. *Sci. Rep.*, 10, 17465.

Chalopin, Y., Piazza, F., Mayboroda, S., Weisbuch, C., Filoche, M. (2019). Universality of fold-encoded localized vibrations in enzymes. *Sci. Rep.*, 9, 12835.

Criado-Reyes, J., Bizzarri, B.M., García-Ruiz, J.M., Saladino, R., Di Mauro, E. (2021). The role of borosilicate glass in Miller-Urey experiment. *Sci. Rep.*, 11, 21009.

Crick, F.H.C. (1968). The origin of the genetic code. *J. Mol. Biol.*, 38, 367–379.

Crick, F.H.C., Brenner, S., Klug, A., Pieczenik, G. (1976). A speculation on the origin of protein synthesis. *Orig. Life*, 7, 389–397.

Demeshkina, N., Jenner, L., Westhof, E., Yusupov, M., Yusupova, G. (2012). A new understanding of the decoding principle on the ribosome. *Nature*, 484, 256–259.

Dibrova, D.V., Galperin, M.Y., Koonin, E.V., Mulkidjanian, A.Y. (2015). Ancient systems of sodium/potassium homeostasis as predecessors of membrane bioenergetics. *Biochemistry (Moscow)*, 80, 495–516.

Eigen, M. and Schuster, P. (1978). The hypercycle. A principle of natural self-organization. Part C: The realistic hypercycle. *Naturwissenschaften*, 65, 341–369.

Eisinger, J. (1971). Complex formation between transfer RNA's with complementary anticodons. *Biochem. Biophys. Res. Comm.*, 43, 854–861.

Gangurde, R., Kaushik, N., Singh, K., Modak, M.J. (2000). A carboxylate triad is essential for the polymerase activity of *Escherichia coli* DNA polymerase (Klenow fragment). Presence of two catalytic triads at the catalytic center. *J. Biol. Chem.*, 275, 19685–19692.

Grosjean, H. and Westhof, G. (2016). An integrated structure- and energy-based view of the genetic code. *Nucleic Acids Res.*, 44, 8020–8040.

Grosjean, H., Söll, D.G., Crothers, D.M. (1976). Studies on the complex between transfer RNAs with complementary anticodons. I. Origins of enhanced affinity between complementary triplets. *J. Mol. Biol.*, 103, 499–519.

Grosjean, H., de Henau, S., Crothers, D.M. (1978). On the physical basis for ambiguity in genetic coding interactions. *Proc. Natl. Acad. Sci. USA*, 75, 610–614.

Hartman, H. (1975). Speculations on the evolution of the genetic code. *Orig. Life*, 6, 423–427.

Hunter, C.A. (1993). Sequence-dependent DNA structure. The role of base stacking interactions. *J. Mol. Biol.*, 230, 1025–1054.

Jordan, I.K., Kondrashov, F.A., Adzhubei, I.A, Wolf, Y.I., Koonin, E.V., Kondrashov, A.S., Sunyaev, S. (2005). A universal trend of amino acid gain and loss in protein evolution. *Nature*, 433, 633–638.

Jukes, T.H. (1967). Indications of an evolutionary pathway in the amino acid code. *Biochem. Biophys. Res. Commun.*, 27, 573–578.

Komar, A.A. (2021). A code within a code: How codons fine-tune protein folding in the cell. *Biochemistry (Moscow)*, 86, 976–991.

Koonin, E.V. and Novozhilov, A.S. (2017). Origin and evolution of the universal genetic code. *Annu. Rev. Genet.*, 51, 45–62.

Miller, S.L. (1953). Production of amino acids under possible primitive earth conditions. *Science*, 117, 528–529.

Miller, S.L. and Urey, H.C. (1959). Organic compound synthesis on the primitive earth. *Science*, 130, 245–251.

Mulkidjanian, A.Y., Bychkov, A.Y., Dibrova, D.V., Galperin, M.Y., Koonin, E.V. (2012). Origin of first cells at terrestrial, anoxic geothermal fields. *Proc. Natl. Acad. Sci. USA*, 109, E821–E830.

Murphy IV, F.V., and Ramakrishnan, V. (2004). Structure of a purine–purine wobble base pair in the decoding center of the ribosome. *Nat. Struct. Mol. Biol.*, 11, 1251–1252.

Ogle, J.M. and Ramakrishnan, V. (2005). Structural insights into translational fidelity. *Annu. Rev. Biochem.*, 74, 129–177.

Pearce, B.K.D., Pudritz, R.E., Semenov, D.A., Henning T.K. (2017). Origin of the RNA world: The fate of nucleobases in warm little ponds. *Proc. Natl. Acad. Sci. USA*, 114, 11327–11332.

Plumridge, A., Andresen, K., Pollack, L. (2020). Visualizing disordered single-stranded RNA: Connecting sequence, structure and electrostatics. *J. Am. Chem. Soc.*, 142, 109–119.

Protozanova, E., Yakovchuk, P., Frank-Kamenetskii, M.D. (2004). Stacked-unstacked equilibrium at the nick site of DNA. *J. Mol. Biol.*, 342, 775–785.

Rybenkov, V.V., Vologodskii, A.V., Cozzarelli, N.R. (1997). The effect of ionic conditions on DNA helical repeat, effective diameter and free energy of supercoiling. *Nucleic Acids Res.*, 25, 1412–1418.

Saladino, R., Di Mauro, E., García-Ruiz, J.M. (2019). A universal geochemical scenario for formamide condensation and prebiotic chemistry. *Chemistry*, 25, 3181–3189.

SantaLucia Jr., J., (1998). A unified view of polymer, dumbbell, and oligonucleotide DNA nearest-neighbor thermodynamics. *Proc. Natl. Acad. Sci. USA*, 95, 1460–1465.

Schlessinger, A. and Rost, B. (2005). Protein flexibility and rigidity predicted from sequence. *Proteins: Structure Funct. Bioinfor.*, 61, 115–126.

Schultz, S.G. and Solomon, A.K. (1961). Cation transport in *Escherichia coli*. I. Intracellular Na and K concentrations and net cation movement. *J. Gen. Physiol.*, 45, 355–369.

Schulz, E.C., Seiler, M., Zuliani, C., Voight, F., Rybin, V., Pogenberg, V., Mücke, N., Wilmanns, M., Gibson, T.J., Barabas, O. (2017). Intermolecular base-stacking mediates RNA-RNA interactions in a crystal structure of the RNA chaperone Hfq. *Sci. Rep.*, 7, 9903.

Seol, Y., Skinner, G.M., Visscher, K., Buhot, A., Halperin, A. (2007). Homopolymeric RNA reveals single-stranded helices and base-stacking. *Phys. Rev. Lett.*, 98, 158103.

Steitz, T.A. (1998). A mechanism for all polymerases. *Nature*, 391, 231–232.

Travers, A. (2006). The evolution of the genetic code revisited. *Orig. Life Evol. Biosph.*, 36, 549–555.

Travers, A. (2022). *Why DNA?* Cambridge University Press, Cambridge.

Trifonov, E.N. (2000). Consensus temporal order of amino acids and evolution of the triplet code. *Gene*, 261, 139–151.

Trifonov, E.N. (2004). The triplet code from first principles. *J. Biomol. Struct. Dyn.*, 22, 1–11.

Trifonov, E.N. (2009). The origin of the genetic code and of the earliest oligopeptides. *Res. Microbiol.*, 160, 481–486.

Vologodskii, A. and Frank-Kamenetskii, M.D. (2018). DNA melting and energetics of the double helix. *Phys. Life Rev.*, 25, 1–21.

Xia, T., SantaLucia, J., Burkard, M.E., Kierzek, R., Schroeder, S.J., Jiao, X., Cox, C., Turner, D.H. (1998). Thermodynamic parameters for an expanded nearest-neighbor model for formation of RNA duplexes with Watson-Crick base pairs. *Biochemistry*, 37, 14719–14735.

Xu, Y.-C. and Bremer, H. (1997). Winding of the DNA helix by divalent metal ions. *Nucleic Acids Res.*, 25, 4067–4071.

Yarus, M. (1982). Translational efficiency of transfer RNA's: Uses of an extended anticodon. *Science*, 218, 646–652.

Yuan, Z., Zhao, J., Wang, Z.-X. (2005). Flexibility analysis of enzyme active sites by crystallographic temperature factors. *Protein Eng. Des. Selec.*, 16, 109–114.

List of Authors

Bruno Mattia BIZZARRI
Ecological and Biological
Sciences Department
University of Tuscia
Viterbo
Italy

Giovanna COSTANZO
Institute of Molecular Biology
and Pathology
CNR
Rome
Italy

David DEAMER
Biomolecular Engineering
University of California
Santa Cruz
USA

Ernesto DI MAURO
Institute of Molecular Biology
and Pathology
CNR
Rome
Italy

Kunio KAWAMURA
Department of Human
Environmental Studies
Hiroshima Shudo University
Japan

Jean-François LAMBERT
Laboratoire de Réactivité de Surface
LRS, UMR 7197 CNRS
Sorbonne Université
Paris
France

Antonio LAZCANO
Universidad Nacional
Autónoma de México
and
El Colegio Nacional
Mexico City
Mexico

Guillaume LESEIGNEUR
Institut de Chimie de Nice
UMR 7272 CNRS
Université Côte d'Azur
France

Marie-Christine MAUREL
Institut de Systématique, Évolution,
Biodiversité (ISYEB)
École Pratique des Hautes Études
Muséum National d'Histoire Naturelle
Sorbonne Université
Université des Antilles
Paris
France

Uwe MEIERHENRICH
Institut de Chimie de Nice
UMR 7272 CNRS
Université Côte d'Azur
France

Juli PERETÓ
Institute for Integrative
Systems Biology (I2SysBio)
CSIC
and
Department of Biochemistry
and Molecular Biology
Universitat de València
Spain

Juan PÉREZ-MERCADER
Department of Earth and Planetary
Sciences and Origins of Life Initiative
Harvard University
Cambridge
Massachusetts
and
Santa Fe Institute
New Mexico
USA

Raffaele SALADINO
Ecological and Biological
Sciences Department
University of Tuscia
Viterbo
Italy

Louis TER-OVANESSIAN
Laboratoire de Réactivité de Surface
LRS, UMR 7197 CNRS
Sorbonne Université
and
Institut de Systématique, Évolution,
Biodiversité (ISYEB)
École Pratique des Hautes Études
Muséum National d'Histoire Naturelle
Sorbonne Université
Université des Antilles
Paris
France

Andrew TRAVERS
MRC Laboratory of Molecular Biology
Cambridge Biomedical Campus
University of Cambridge
United Kingdom

Juan VLADILO
INAF – Trieste Astronomical
Observatory
Italy

Index

A, B

abiogenesis, 20–23
abiotic life, 135, 136, 138, 145, 149, 151, 153
adsorption, 194, 196–202, 206
Albery, 87
amino acids, 194, 197, 199, 200, 202, 203, 205, 208
    recruitment, 220
    synthesis from formamide, 226
ammonia, 6, 7
Andromeda, 19
Archean, 22
autocatalytic cycle, 85
autopoiesis, 2
base-pair wobble, 220
Big Bang, 11, 12
binary systems, 19
biosignature, 3, 23
    atmospheric, 23

C

carbon (C), 6, 11, 12, 15, 21
carbonaceous chondrite(s) (CC), 14, 21
Cárdenas, 92
catalysis
    enzymatic, 195, 198, 204–207, 209
    heterogeneous, 197, 204–207, 209, 210
catalytic molecules, 4–8
chemical
    disequilibrium, 3, 23
    evolution, 138
chemomimesis, 59
chiral amplification, 38, 39
chirality, 197
circular dichroism, 39, 40
circularly polarized light (CPL), 34, 39, 40, 42
clay minerals, 194, 196, 198, 199, 205
coacervates, 121–123, 127, 129
comet 67P/Churyumov-Gerasimenko, 42–44
COmetary Sampling And Composition (COSAC), 43–46
compartments, 3, 5, 6, 20
competition in early life, 139, 148, 155, 156
complex organic molecules (COMs), 12

complexes, 196, 197
Cornish-Bowden, 87, 91, 92
cosmic microwave background (CMB), 11

D, E

Darwin, 79–82, 90
Dayhoff, 81, 85
de Duve, 82, 83, 85
de Lorenzo, 82, 93
deep ocean, 171, 172, 174, 176, 185
design, 79, 80
electron-to-proton mass ratio, 10
emission spectroscopy, 24
enantiomeric excess (*ee*), 21, 22
energy sources, 55, 66–68
entropy, 3, 5
enzyme, 163, 172, 181–183
evolution of genetic code, 217, 218
exogenous delivery, 21, 23
extrasolar planet (exoplanet), 15, 16, 18, 23, 24

F, G

fine structure constant, 10
Fischer, 34–36
flow reactor, 176–179
formamide, 6, 12, 55–57, 62, 67
free energy, 195, 202, 208
galactic habitable zone (GHZ), 19
genetic molecules, 4, 6–9
geochemistry/geochemical, 194, 195, 198, 199, 204, 205, 207, 209, 210
giant planets, 9, 15, 16
global circulation model (GCM), 17

H

habitability, 3, 8–10, 13, 16–20, 24
habitable zone (HZ), 1, 17, 19

Haeckel, 117, 118, 122, 130
heterotrophic
  hypothesis, 120–122, 128, 129
  origin of life, 23
hydrogen
  bond (HB/H bond), 4–8, 196–198
  bonding, 167, 171–173, 182
hydrolysis, 171
hydrothermal, 165–168, 170, 171, 173–179, 181–186
  vent, 22

I

icy moons, 23
infinite regression, 91, 92
information flow, 163, 164
insoluble organic material (IOM), 14
instellation, 10, 16–19
instrumentation, 175
interactions
  electrostatic, 172, 182
  hydrophobic, 171–173, 182
intermolecular interactions, 4, 5
interstellar
  dust grains, 40, 42
  medium (ISM), 8, 12
intramolecular bonding, 5
ionizing radiation, 9
iron(Fe), 199, 205

J, K, L

Jacob, 79, 81, 82, 85, 90, 92
Jupiter, 9, 23
kinetics, 179, 204, 205, 210
kissing loops, 217, 219–221
Knowles, 87
Krishnamurthy, 84
Lazcano, A., 82, 83, 87

life
  generalized, 136, 138, 142, 152–154
  liposomes, 127, 128

M

main-sequence stars, 17, 18
Mars, 17
  Organic Molecules Analyzer (MOMA), 45, 46
metabolic closure, 91
metabolism, 193–195
metallicity, 12, 15, 19
methane, 6, 7, 17
Milky Way, 11, 12, 15, 19
Miller, 82, 83, 87
  electric discharge experiments, 224, 226
minerals, 56–64, 168, 169, 171, 174, 176–178, 183–185, 193–196, 198, 199, 202–210
molecular
  bricolage, 81, 82, 93
  chirality, 21, 23
  clouds, 8, 11, 12, 14
  evolution, 55, 61
  tinkering, 85
moonlighting, 91, 92
Morange, 82, 87, 91
Muller, 121–123

N, O

nitrogen (N), 21
nucleobases, 206, 209
nucleosynthesis, 11, 12
nucleotides, 194, 202, 203, 208
ocean worlds, 23, 46
Oparin, 118–129

optical rotation, 33, 34, 39
origin(s) of life (*see also* heterotrophic), 117, 118, 120–122, 124, 126–130, 136, 137, 139, 140, 143
  autotrophic, 22
origin and evolution of life, 111
oxygen (O), 14

P

parity-violating energy difference (PVED), 39
Pasteur, 33, 34
peptide, 164, 171, 172, 174, 175, 179, 182
perfect enzyme, 87
phosphorus (P), 7
planetary
  climate, 10, 16, 17
  systems, 12, 13, 16
planetesimals, 13–15, 21
polymer chemistry, 135, 140, 141, 144, 150, 152, 157
polymerization, 194, 199–202
Pop I, 12, 15
Pop II, 12, 15
Pop III, 11, 15
potassium (K), 22
prebiotic
  chemistry, 31, 55–58, 60, 61, 64, 65, 68
  synthesis, 127
precellular evolution, 123, 126
pressure, 7–9, 164–167, 170–178, 183–185
primitive Earth, 120, 121, 123–125, 127, 128
promiscuity, 85–87, 91, 92
protocells, 137, 138, 150–152, 154, 156

protocellular systems, 126
protometabolism, 82–85
protoplanetary disk, 13, 14

R

racemic, 33, 34, 37
racemization, 36, 37, 45
RAFT polymerization, 141–143, 155
recombination, 11
regioselectivity, 65–67
ribozyme, 164, 168–170, 172, 177–179, 182–184
RNA, 55, 57, 60, 61, 65, 68
   -like molecules, 99, 111
   base modification, 220, 222
   base stacking, 221, 222, 226
   circular, 109
   duplex melting energy, 221
   viral, 105, 107, 111
   world, 99, 110, 111, 163–165, 168, 175, 178
rocky planets, 1, 10, 12, 14–17, 19
Rosen, 91, 92

S

selectivity, 197, 198, 206, 207, 209
self-assembly self-reproduction, 135, 136, 140–144, 152, 155, 156
self-booting, 150, 153,
self-replication, 163, 168, 182
serpentinization, 199
serpentization, 226
silica (SiO_2), 196–198, 202, 204, 208, 209
solar-type stars, 13, 17, 18, 24
solubility, 173

space mission
   DragonFly, 46
   ExoMars, 31, 43, 45, 46
   Rosetta, 31, 42–46
Springsteen, 84
stereoselectivity, 65, 66
subcellular systems, 126, 127
sulphur (S), 7
Super-Earths, 16
surfaces, 193–199, 201–204, 207–210
synthetic
   biology, 82, 92, 93
   life, 135, 140, 143, 149, 152, 154
systems chemistry, 82–85

T

Tawfik, 79, 86, 88, 89, 93
temperature, 7–11, 13, 17, 22, 165–168, 170–176, 178–185
trajectory, 168, 183, 184
thermodynamics, 170, 171, 174, 178, 181, 198, 199, 204, 210
transmission spectroscopy, 24
triple point, 7, 8
tRNA structure, 219
two-gene hypothesis, 168

V, W, Z

van der Waals (vdW), 4, 5
Venus, 17
viroids, 99, 108–111
viruses, 99–104, 106–109, 111
water, 6–8, 13, 14, 16, 17, 20–22, 165–168, 170–176, 182, 185
wetting-and-drying, 202, 203
zeolites, 196, 205, 209

Printed and bound by CPI Group (UK) Ltd, Croydon, CR0 4YY
22/01/2024